A Graduate Course in Probability

Probability and Mathematical Statistics

A Series of Monographs and Textbooks

Edited by

Z. W. Birnbaum
University of Washington
Seattle, Washington

E. Lukacs
Catholic University
Washington, D.C.

---

1. Thomas Ferguson. Mathematical Statistics: A Decision Theoretic Approach. 1967

2. Howard Tucker. A Graduate Course in Probability. 1967

In preparation

K. R. Parthasarathy. Probability Measures on Metric Spaces

Preface

This book is based on a course in probability that is given to graduate students in the Department of Mathematics at the University of California, Riverside. Its purpose is to define one possible course in probability theory that might be given at the graduate level. The prerequisite for this text is a knowledge of real analysis or measure theory. This book does not contain a review of measure theory, but whenever a theorem from measure theory is used, it is usually stated in full. In particular, the student or reader should be familiar with the Lebesgue integral over abstract spaces and its properties and, in particular, the Lebesgue dominated convergence theorem, Fubini's theorem, the Radon-Nikodym theorem, Egorov's theorem, the monotone convergence theorem, and the theorem on unique extension of a sigma-finite measure from an algebra to the sigma-algebra generated by it. No previous knowledge of probability is needed. This book otherwise is self-contained and consists of exactly enough material for a one-year course (meeting three hours per week) in probability. An instructor wishing to create time for special topics he wishes to lecture on could omit Sections 5.5 and 5.6 and any block of sections at the end of Chapter 8.

I feel that this book could serve as a text for a one-year graduate course in probability given in a mathematics program, preferably for students in their second year of graduate work. It is especially suited for a rigorous graduate course in probability needed for a mathematical statistics program. In addition, the first four chapters could be used for the first quarter of a one-year course in mathematical statistics given to mathematically mature graduate students who have never before studied probability and statistics (but who know real analysis). Last, because I have chosen to include as much explanation for the proof of each theorem as possible, and because I have resisted the urge to include anything not immediately relevant, I feel that this book could be read with profit by any mature mathematician who wishes in a reasonably short time to become acquainted with some of the basic theorems in probability.

The selection of material reflects my taste for such a course. I have attempted here what I consider a proper balance between measure–theoretic aspects of probability (for example, strong limit laws, martingale theory, conditional expectation, stochastic processes) and distributional aspects (distribution functions, characteristic functions, central limit theorems). Many important topics are omitted; for example, there is no mention of Markov chains, nor of ergodic theory, nor of Poisson processes. The material presented does not wander along any scenic byways. Rather, I was interested in traveling the shortest route to certain theorems I wished to present. These theorems are the strong limit theorems of Chapter 5, the general limit theorem (Theorem 3) given in Section 6.5, the special limit theorems in Section 6.6, and the theorems given in Sections 7.3, 8.3, and 8.4. Material contained in all other sections and chapters of this book are results needed to provide a minimal rigorous pathway to these theorems.

My primary aim in this book was to present some of the basic theorems of analytic probability theory in a cohesive manner. I did not wish necessarily to present these in their most general and extended form. Rather, I wished to render statements of these theorems as simple as possible in order that they be easy to remember and so that the essential idea behind each proof is visible. Also I wished to put these theorems in the form most frequently referred to in research papers. Thus the presentation given here does not give the last word on any of the results. For the last word, an approach can be made through the monumental works of J. L. Doob, M. Loève, and W. Feller listed in the suggested reading section at the end of the book. I wish to call attention to the problems at the end of each section. I feel that every problem should be assigned to students taking a course in which this book is the text. The problems were designed not to increase the contents of the book but to enhance those contents already there. I feel I achieved my best success in these efforts with the problems designed for Section 6.5.

I wish to acknowledge my gratitude to a few of the number of people whose collective help both contributed to many smoother proofs and in several cases kept me from committing grievous errors. They are Howard H. Stratton, Jr., Lynn G. Gref, J. David Mason, William B. Stelwagon, and Lambert H. Koopmans. The writing of this book was supported in part by the Air Force Office of Scientific Research, Grant Number AF-AFOSR 851-65; this support is very much appreciated. Mrs. Jane Scully deserves my gratitude for her accurate and speedy typing of the manuscript. And, finally, I acknowledge my gratitude to Academic Press Inc. for their cooperation in this endeavor.

Howard G. Tucker

Riverside, California

Contents

Preface vii

CHAPTER 1 **Probability Spaces**

1.1   Sigma Fields   1
1.2   Probability Measures   5
1.3   Random Variables   9

CHAPTER 2 **Probability Distributions**

2.1   Univariate Distribution Functions   15
2.2   Multivariate Distribution Functions   23
2.3   Distribution of a Set of Infinitely Many Random Variables   29
2.4   Expectation   34
2.5   Characteristic Functions   41

CHAPTER 3 **Stochastic Independence**

3.1   Independent Events   57
3.2   Independent Random Variables   61
3.3   The Zero-One Law   70

CHAPTER 4   **Basic Limiting Operations**

4.1   Convergence of Distribution Functions............................. 77
4.2   The Continuity Theorem... 88
4.3   Refinements of the Continuity Theorem for Nonvanishing
     Characteristic Functions... 92
4.4   The Four Types of Convergence: Almost Sure, in Law, in
     Probability, and in rth Mean................................... 99

CHAPTER 5   **Strong Limit Theorems for Independent
             Random Variables**

5.1   Almost Sure Convergence of Series of Independent Random
     Variables.. 107
5.2   Proof that Convergence in Law of a Series of Independent Random
     Variables Implies Almost Sure Convergence...................... 115
5.3   The Strong Law of Large Numbers............................... 122
5.4   The Glivenko-Cantelli Theorem................................. 126
5.5   Inequalities for the Law of the Iterated Logarithm................. 129
5.6   The Law of the Iterated Logarithm.............................. 137

CHAPTER 6   **The Central Limit Theorem**

6.1   Infinitely Divisible Distributions................................ 147
6.2   Canonical Representation of Infinitely Divisible Characteristic
     Functions.. 153
6.3   Convergence of Infinitely Divisible Distribution Functions........... 161
6.4   Infinitesimal Systems of Random Variables....................... 167
6.5   The General Limit Theorem for Sequences of Sums of Independent
     Random Variables.. 177
6.6   Convergence to the Normal and Poisson Distributions............... 194

CHAPTER 7   **Conditional Expectation and Martingale
             Theory**

7.1   Conditional Expectation... 209
7.2   Martingales and Submartingales................................. 220
7.3   Martingale and Submartingale Convergence Theorems.............. 229
7.4   Brownian Motion... 239

CHAPTER 8   **An Introduction to Stochastic Processes
             and, in Particular, Brownian Motion**

8.1   Probability Measures over Function Spaces........................ 245
8.2   Separable Stochastic Processes.................................. 249

8.3   Continuity and Nonrectifiability of Almost All Sample Functions
        of Separable Brownian Motion................................   254
8.4   The Law of the Iterated Logarithm for Separable Brownian Motion....   265

Suggested Reading..   270

Index   271

CHAPTER 1

Probability Spaces

1.1 Sigma Fields

The very beginning of our considerations deals with a space or set Ω. The set Ω consists of elements or points, each of which is an individual outcome of a game or experiment or other random phenomenon under consideration. If the game is a toss of a die, then Ω consists of six points. Sometimes the points of Ω are the set of real numbers, sometimes they are a set of functions, and sometimes they are the set of all points in Euclidean n-dimensional space $E^{(n)}$. Each point or element ω in Ω will be referred to as an *elementary event*, and Ω will be called the *fundamental probability set*, or the *sure event*.

In everything that follows we shall denote subsets of a fundamental probability set Ω by upper case letters from the beginning of the English alphabet. By $\omega \in A$ we mean that ω is an elementary event in A. As usual, $A \cup B$ denotes the union of A and B, and $A \cap B$ or AB denotes the intersection of A and B. The symbol ϕ denotes the empty set. The complement of A in Ω is denoted by A^c, and $A \setminus B$ denotes the set of points in A which are not in B, that is, $A \setminus B = AB^c$. If $\{B_n\}$ is a collection of subsets, then $\cup B_n$ denotes their union and $\cap B_n$ denotes their intersection.

Associated with a *sure event* Ω is a nonempty set of subsets of Ω, denoted by $\mathfrak{A}$, called a *sigma field* (or *sigma algebra*) of subsets of Ω.

1

Definition: A set of subsets $\mathcal{A}$ of Ω is called a sigma field if

(a) for every $A \in \mathcal{A}$, then also $A^c \in \mathcal{A}$,
(b) if $A_1, A_2, \cdots, A_n, \cdots$ is a countable sequence of elements of $\mathcal{A}$, then $\cup A_n \in \mathcal{A}$, and
(c) $\phi \in \mathcal{A}$.

The elements or members of a sigma field $\mathcal{A}$ of subsets of Ω are called *events*, and much of the set-theoretic terminology is translated into the terminology of events. If the elementary event ω occurs and if $\omega \in A$, then we say that the event A occurs. If $A \in \mathcal{A}$ and $B \in \mathcal{A}$, then AB or $A \cap B$ means that both the events A and B occur, and $A \cup B$ means the event that at least one of these two events occurs. The complement A^c of A means the event that A does not occur, and $A \setminus B$ means that A occurs and B does not occur. Since we call Ω the *sure event*, we shall refer to ϕ as the *impossible event*. If A and B are disjoint events, that is, if they have no elementary events in common or $AB = \phi$, then we say that A and B are incompatible; that is, they cannot both occur. If $A \subset B$, that is, if every elementary event in A is also in B, we say that the occurrence of the event A implies the occurrence of the event B, or A implies B.

Theorem 1. $\Omega \in \mathcal{A}$.

Proof: Referring to the definition of $\mathcal{A}$, $\phi \in \mathcal{A}$, and hence $\phi^c = \Omega \in \mathcal{A}$.

Theorem 2. If $\{A_n\}$ is any countable sequence of events in $\mathcal{A}$, then $\cap A_n \in \mathcal{A}$.

Proof: By the definition of $\mathcal{A}$, $A_n{}^c \in \mathcal{A}$, and thus $\cup A_n{}^c \in \mathcal{A}$. Again by the definition of $\mathcal{A}$, $(\cup A_n{}^c)^c \in \mathcal{A}$, and by the DeMorgan formula, $(\cup A_n{}^c)^c = \cap A_n$, which completes the proof.

Definition: If $\{A_n\}$ is a (denumerable) sequence of sets, then we define $\lim \sup A_n$ and $\lim \inf A_n$ by

$$\lim \sup A_n = \bigcap_{n=1}^{\infty} \bigcup_{k=n}^{\infty} A_k$$

and

$$\lim \inf A_n = \bigcup_{n=1}^{\infty} \bigcap_{k=n}^{\infty} A_k.$$

If $\lim \sup A_n = \lim \inf A_n$, then we refer to this set by $\lim A_n$, and if we denote $\lim A_n$ by A, then we write $A_n \to A$.

Theorem 3. If $\{A_n\}$ is a sequence of events, then $\lim \sup A_n \in \mathcal{C}$ and $\lim \inf A_n \in \mathcal{C}$.

Proof: This is an immediate consequence of the definition and Theorems 1 and 2.

The event $\lim \sup A_n$ means the event that infinitely many of the events A_n occur, or A_n occurs "infinitely often." This is because a point (or elementary event) is in $\lim \sup A_n$ if and only if it is in infinitely many of the A_n. The event $\lim \inf A_n$ means the event that all but a finite number of the events in $\{A_n\}$ occur, or A_n occurs "almost always." This is because a point is in $\lim \inf A_n$ if and only if it is in all but a finite number of the A_n.

The following redundant definition is stated in order to avoid possible confusion.

Definition: If $\{\mathcal{C}_\lambda, \lambda \in \Lambda\}$ is a collection of sets of subsets of Ω, then $\bigcap_{\lambda \in \Lambda} \mathcal{C}_\lambda$ denotes the set of subsets of Ω, each of which belongs to every $\mathcal{C}_\lambda$.

Theorem 4. If $\{\mathcal{C}_\lambda, \lambda \in \Lambda\}$ is a collection of sigma fields of subsets of Ω, then $\bigcap_{\lambda \in \Lambda} \mathcal{C}_\lambda$ is a sigma field.

Proof: The proof is immediate upon verifying the three requirements of a sigma field.

Definition: Let $\mathcal{C}$ be a collection of subsets of Ω. By the *smallest sigma field containing* $\mathcal{C}$ or the *sigma field generated by* $\mathcal{C}$, which we shall denote by $\sigma\{\mathcal{C}\}$ or $\sigma(\mathcal{C})$, we mean a sigma field of subsets of Ω such that if $\mathcal{B}$ is a sigma field of subsets containing $\mathcal{C}$, then $\mathcal{C} \subset \sigma\{\mathcal{C}\} \subset \mathcal{B}$.

We have just defined an object. We must now prove that it exists and is unique.

Theorem 5. If $\mathcal{C}$ is a collection of subsets of Ω, there exists one and only one sigma field generated by $\mathcal{C}$.

Proof: Let $\{\mathcal{C}_\lambda\}$ be the set of all sigma fields of subsets of Ω such that $\mathcal{C} \subset \mathcal{C}_\lambda$ for all λ. There does exist at least one such $\mathcal{C}_\lambda$ in this set, namely, the set of all subsets of Ω. Now $\mathcal{C} \subset \cap\mathcal{C}_\lambda$, and $\cap\mathcal{C}_\lambda$ is by Theorem 4 a sigma field. Denote $\cap\mathcal{C}_\lambda = \sigma\{\mathcal{C}\}$, and let $\mathcal{B}$ be any sigma field such that $\mathcal{C} \subset \mathcal{B}$. Then $\mathcal{B}$ is an $\mathcal{C}_\lambda$ for some λ, and $\sigma\{\mathcal{C}\} \subset \mathcal{B}$, thus proving the theorem.

The following remark should be kept in mind. If $\{\mathcal{C}_n\}$ is a sequence of sigma fields, then $\cap\mathcal{C}_n$ is a sigma field by Theorem 4. Even if $\mathcal{C}_n \subset \mathcal{C}_{n+1}$ for all n, however, it is not necessarily true that $\cup\mathcal{C}_n$ is a sigma field. In this case, requirement (b) of the definition cannot be verified.

EXERCISES

1. Let $\Omega = (-\infty, +\infty)$, and let $\mathcal{C}_n$ be the sigma field generated by the subsets $[0, 1)$, $[1, 2)$, $\cdots$, $[n - 1, n)$. Prove: (a) $\mathcal{C}_n \subset \mathcal{C}_{n+1}$ for all n, and (b) $\cup_{n=1}^{\infty}\mathcal{C}_n$ is *not* a sigma field.

2. In Problem 1, what are the subsets of Ω which are elements of $\mathcal{C}_2$?

3. Let $\{A_n\}$ be a sequence of events. Define B_m to be the event that the first among the events A_1, A_2, $\cdots$ that occurs is A_m. (a) Express B_m in terms of A_1, A_2, $\cdots$, A_m. (b) Prove that $\{B_m\}$ are disjoint. (c) $\cup_{m=1}^{\infty}B_m = ?$

4. Prove that $\lim\inf A_n \subset \lim\sup A_n$.

5. Write in terms of set-theoretic operations: exactly two of the events A_1, A_2, A_3, A_4 occur.

6. Prove: if $A_n \subset A_{n+1}$ for all n, then $\lim\sup A_n = \lim\inf A_n = \lim A_n$.

7. Prove in two ways that if $A_n \supset A_{n+1}$ for all n, then $\lim\sup A_n = \lim\inf A_n = \lim A_n$.

8. Prove Theorem 3.

9. Prove Theorem 4.

10. Let $\{A_n\}$ be a sequence of events in Ω. Prove that $\cup_{n=1}^{\infty}A_n = \cup_{n=1}^{\infty}B_n$, where $B_1 = A_1$ and $B_n = A_1^c \cap \cdots \cap A_{n-1}^c \cap A_n$ for $n \geq 2$, and prove that the events $\{B_n\}$ are disjoint.

11. Let $\{a_n\}$ be a sequence of real numbers, and let $A_n = (-\infty, a_n]$. Prove that

$$\text{l.u.b. } \limsup A_n = \limsup a_n$$

and

$$\text{l.u.b. } \liminf A_n = \liminf a_n.$$

12. Let $\mathcal{C}_1$, $\mathcal{C}_2$ be two sets of subsets of Ω, and assume that $\mathcal{C}_1 \subset \mathcal{C}_2$. Prove that $\sigma\{\mathcal{C}_1\} \subset \sigma\{\mathcal{C}_2\}$.

1.2 Probability Measures

In the previous section we introduced the notions of fundamental probability set, or sure event, and sigma field $\mathcal{C}$ of events in Ω. In this section the notions of probability and conditional probability are introduced.

Definition: A probability P is a normed measure over a measurable space $(\Omega, \mathcal{C})$; that is, P is a real-valued function which assigns to every $A \in \mathcal{C}$ a number $P(A)$ such that

(a) $P(A) \geq 0$ for every $A \in \mathcal{C}$,
(b) $P(\Omega) = 1$, and
(c) if $\{A_n\}$ is any denumerable union of disjoint events, then

$$P\left(\bigcup_{n=1}^{\infty} A_n \right) = \sum_{n=1}^{\infty} P(A_n).$$

One refers to $P(A)$ as "the probability of (the event) A." From here on, whenever we speak of events and their probabilities it should be understood that a silent reference is made to some fixed fundamental probability space, a sigma field of events, and a probability measure. There are a number of immediate consequences of the definition of a probability.

Theorem 1. $P(\phi) = 0$.

Proof: Denote $A_n = \phi$ for $n = 1, 2, \cdots$. The conclusion follows from the fact that $\phi = \bigcup_{n=1}^{\infty} A_n$ and from (c) in the definition.

Theorem 2. If $A_1, \cdots, A_n$ are any n disjoint events, then

$$P(A_1 \cup \cdots \cup A_n) = \sum_{k=1}^{n} P(A_k).$$

Proof: Let $\phi = A_{n+1} = A_{n+2} = \cdots$. By (c) in the above definition and by Theorem 1,

$$P(A_1 \cup \cdots \cup A_n) = P(\bigcup_{k=1}^{\infty} A_k) = \sum_{k=1}^{\infty} P(A_k) = \sum_{k=1}^{n} P(A_k),$$

which proves the assertion.

Theorem 3. If A and B are events, and if $A \subset B$, then $P(A) \leq P(B)$.

Proof: Since $B = A \cup A^c B$, and since A and $A^c B$ are disjoint, then by Theorem 2 and by (a) in the above definition we have

$$P(B) = P(A) + P(A^c B) \geq P(A),$$

which yields the desired inequality.

Corollary to Theorem 3. For every $A \in \mathfrak{a}$, $P(A) \leq 1$.

Proof: Since $A \in \mathfrak{a}$ implies that $A \subset \Omega$, and since $P(\Omega) = 1$, then by Theorem 3, $P(A) \leq P(\Omega) = 1$.

Theorem 4 (Boole's Inequality). If $\{A_n\}$ is a countable sequence of events, then

$$P(\bigcup_{n=1}^{\infty} A_n) \leq \sum_{n=1}^{\infty} P(A_n).$$

Proof: By Problem 10 in Section 1.1, $\bigcup_{n=1}^{\infty} A_n = \bigcup_{n=1}^{\infty} B_n$, where $B_1 = A_1$ and $B_n = A_1^c \cdots A_{n-1}^c A_n$ for $n \geq 2$. Since the B_n are disjoint, then

$$P(\bigcup_{n=1}^{\infty} A_n) = \sum_{n=1}^{\infty} P(B_n).$$

However, $B_n \subset A_n$ for every n, and so by Theorem 3, $P(B_n) \leq P(A_n)$, which yields the conclusion of the theorem.

Theorem 5. For every event A, $P(A^c) = 1 - P(A)$.

Proof: Since $\Omega = A \cup A^c$, we obtain from Theorem 2 that $1 = P(\Omega) = P(A) + P(A^c)$, which is equivalent to the conclusion.

The triplet $(\Omega, \mathcal{Q}, P)$ will be referred to as a *probability space*. If Ω is a countable set, then $\mathcal{Q}$ is usually the set of all subsets of Ω, and there is no difficulty in defining a probability measure P over $\mathcal{Q}$ which has the countable additivity property (c) in the above definition. If Ω is uncountable, however, then $\mathcal{Q}$ cannot in general be the set of all subsets.

Once one has one probability defined over $(\Omega, \mathcal{Q})$, then one can define other probabilities that are called conditional probabilities.

Definition: If $A \in \mathcal{Q}$ and $B \in \mathcal{Q}$ and if $P(B) > 0$, then the conditional probability of A given B, $P(A \mid B)$, is defined by $P(A \mid B) = P(AB)/P(B)$.

An interpretation of $P(A \mid B)$ is that it is the probability of the event A occurring if one knows that B occurs.

Theorem 6. If $B \in \mathcal{Q}$ and $P(B) > 0$, then $P(\cdot \mid B)$, as a function over $\mathcal{Q}$, is a probability; that is,

(a) $P(A \mid B) \geq 0$ for every $A \in \mathcal{Q}$,
(b) $P(\Omega \mid B) = 1$, and
(c) $P(\cup_{n=1}^{\infty} A_n \mid B) = \sum_{n=1}^{\infty} P(A_n \mid B)$ for every denumerable sequence of disjoint events $\{A_n\}$ in $\mathcal{Q}$.

Proof: One can easily verify (a), (b), and (c) by direct application of the definitions of probability and conditional probability.

Two very important and useful properties of conditional probabilities are the following two theorems.

Theorem 7 (Multiplication Rule). For every $n + 1$ events A_0, $A_1, \cdots, A_n$ for which $P(A_0 A_1 \cdots A_{n-1}) > 0$, we have

$$P(A_0 A_1 \cdots A_n) = P(A_0) P(A_1 \mid A_0) \cdots P(A_n \mid A_0 A_1 \cdots A_{n-1}).$$

Proof: Since $A_0A_1\cdots A_{n-1} \subset A_0A_1\cdots A_{n-2} \subset \cdots \subset A_0$, then

$$0 < P(A_0A_1\cdots A_{n-1}) \leq \cdots \leq P(A_0),$$

and consequently all the conditional probabilities involved in the statement of the theorem are well defined. The conclusion is clearly true for $n = 1$ by direct application of the definition of conditional probability. The rest of the proof is an easy application of mathematical induction.

Theorem 8 (Theorem of Total Probabilities). If $P(\cup_{n=1}^N B_n) = 1$, where $\{B_n\}$ are a finite or denumerable sequence of disjoint events, if $P(B_n) > 0$ for every n, and if $A \in \mathcal{C}$, then

$$P(A) = \sum_{n=1}^N P(A \mid B_n)P(B_n).$$

Proof: Since $P((\cup_{n=1}^N B_n)^c) = 0$, then

$$P(A) = P(A \cap (\bigcup_{n=1}^N B_n)) + P(A \cap (\bigcup_{n=1}^N B_n)^c)$$

$$= P(\bigcup_{n=1}^N AB_n) = \sum_{n=1}^N P(AB_n)$$

$$= \sum_{n=1}^N P(A \mid B_n)P(B_n),$$

which concludes the proof.

A commonly used probability space for the construction of examples and counterexamples in probability theory is the *unit-interval probability space* which we define as follows: Let $\Omega = [0, 1]$, let $\mathcal{C}$ be the sigma field of all Lebesgue-measurable subsets of Ω, and P is the ordinary Lebesgue measure defined over $[0, 1]$. The experiment or game which gives rise to such a probability space is that of selecting a real number "at random" between 0 and 1.

EXERCISES

1. Prove that $P(A \cup B) = P(A) + P(B) - P(AB)$.
2. Prove Bayes' Rule: If $P(\cup_{n=1}^N B_n) = 1$, where $\{B_n\}$ are a finite or a denumerable sequence of disjoint events such that $P(B_n) > 0$ for

all n, and if $A \in \alpha$ and $P(A) > 0$, then for every k

$$P(B_k \mid A) = P(A \mid B_k)P(B_k) / \sum_{n=1}^{N} P(A \mid B_n)P(B_n).$$

3. Prove that if $A_n \to A$, then $P(A_n) \to P(A)$.

4. For A, B in α, define $\rho(A, B) = P(AB^c) + P(A^cB)$. Prove that (α, ρ) is a pseudometric space.

5. Complete the proofs of Theorems 6 and 7.

6. Suppose $\Omega = (-\infty, +\infty)$, and let α be the set of all subsets of $(-\infty, +\infty)$. For every $A \in \alpha$ define $P(A) = \sum \{2^{-n} \mid n \text{ is a positive integer and } n \in A\}$. Is P a probability?

1.3 Random Variables

In the first two sections we developed the concept of a probability space. This mathematical model completely describes the outcome of an experiment or game. However, in order to answer questions concerning the experiment one would have to be able to observe an ω selected at random according to the probability measure P. Usually all that is needed is that a function of ω be observed. Such a function is sometimes called a random variable. This section is an introduction to the concept of random variable.

Definition: A random variable X is a real-valued function whose domain is Ω and which is α-measurable, that is, for every real number x, $\{\omega \in \Omega \mid X(\omega) \leq x\} \in \alpha$.

The expression in the curly brackets above denotes the set of elementary events ω in Ω for which $X(\omega) \leq x$. We shorten this notation to $[X \leq x]$; that is, we denote

$$[X \leq x] = \{\omega \in \Omega \mid X(\omega) \leq x\}.$$

Similarly we denote

$$[X < x] = \{\omega \in \Omega \mid X(\omega) < x\},$$

and, in general, for any Borel set B we denote

$$[X \in B] = \{\omega \in \Omega \mid X(\omega) \in B\}.$$

Definition: If f is a function with domain $\mathfrak{X}$ and range $\mathcal{Y}$, and if $D \subset \mathcal{Y}$, then we define

$$f^{-1}(D) = \{x \in \mathfrak{X} \mid f(x) \in D\}.$$

If $\mathfrak{D}$ is a collection of subsets $\{D\}$ of $\mathcal{Y}$, then we define

$$f^{-1}(\mathfrak{D}) = \{f^{-1}(D) \mid D \in \mathfrak{D}\}.$$

Proposition 1. If f is a function whose domain is $\mathfrak{X}$ and whose range is in $\mathcal{Y}$, and if $\mathcal{C}$ is a nonempty collection of subsets of $\mathcal{Y}$, then

$$\sigma\{f^{-1}(\mathcal{C})\} = f^{-1}(\sigma\{\mathcal{C}\}).$$

(Recall that $\sigma\{\mathcal{C}\}$ denotes the sigma field generated by $\mathcal{C}$.)

Proof: We repeatedly use the fact that for any collection of subsets $\{C_\lambda\}$ of $\mathcal{Y}$, $f^{-1}(\cup C_\lambda) = \cup f^{-1}(C_\lambda)$ and $f^{-1}(\cap C_\lambda) = \cap f^{-1}(C_\lambda)$. Denote $\mathcal{C}_1 = \{C \mid C \in \mathcal{C} \text{ or } C^c \in \mathcal{C}\}$, and for each countable ordinal number $\mu > 1$, denote $\mathcal{C}_\mu$ as the set of all countable unions and all countable intersections of elements in $\cup_{\nu < \mu} \mathcal{C}_\nu$. If M denotes the set of all countable ordinals, then $\cup_{\mu \in M} \mathcal{C}_\mu$ is a sigma field, and moreover $\sigma\{\mathcal{C}\} = \cup_{\mu \in M} \mathcal{C}_\mu$. Hence by the above remark, $f^{-1}(\sigma\{\mathcal{C}\}) = \cup_{\mu \in M} f^{-1}(\mathcal{C}_\mu)$. Again by the above remark $f^{-1}(\mathcal{C}_\mu)$ is the set of all countable unions and intersections of elements in $\cup_{\nu < \mu} f^{-1}(\mathcal{C}_\nu)$. Hence $\cup_{\mu \in M} f^{-1}(\mathcal{C}_\mu)$ is a sigma field and in fact is the sigma field generated by $f^{-1}(\mathcal{C})$. Hence $f^{-1}(\sigma\{\mathcal{C}\}) = \sigma\{f^{-1}(\mathcal{C})\}$, which is the conclusion.

Definition: An n-dimensional random variable or an n-dimensional random vector or a vector random variable, $\mathbf{X} = (X_1, \cdots, X_n)$, is a function whose domain is Ω, whose range is in Euclidean n-space $E^{(n)}$, and such that for every Borel-measurable subset $B \subset E^{(n)}$,

$$\{\omega \in \Omega \mid (X_1(\omega), \cdots, X_n(\omega)) \in B\} \in \mathcal{A}.$$

As in the one-dimensional case, so in the n-dimensional case we use the notation

$$[\mathbf{X} \in B] = \{\omega \in \Omega \mid (X_1(\omega), \cdots, X_n(\omega)) \in B\},$$

where $\mathbf{X} = (X_1, \cdots, X_n)$.

Proposition 2. If $f:E^{(n)} \to E^{(m)}$ is a Borel measurable function, that is, if $B \subset E^{(m)}$ is a Borel set, then $f^{-1}(B) \subset E^{(n)}$ is a Borel set, and if **X** is an n-dimensional random variable $(n \geq 1)$, then $f(\mathbf{X})$ is an m-dimensional random variable.

Proof: Let $B \subset E^{(m)}$ be a Borel set; we must prove that $[\, f(\mathbf{X}) \in B\,] \in \mathcal{Q}$. But

$$[\, f(\mathbf{X}) \in B\,] = [\,\mathbf{X} \in f^{-1}(B)\,] \in \mathcal{Q}$$

since $f^{-1}(B)$ is a Borel set, which concludes the proof.

Proposition 3. If $X_1, \cdots, X_n$ are n random variables, then $\mathbf{X} = (X_1, \cdots, X_n)$ is an n-dimensional random variable.

Proof: By hypothesis and because of the properties of $\mathcal{Q}$,

$$\bigcap_{i=1}^{n} [X_i \leq x_i] = \Big[\mathbf{X} \in \underset{i=1}{\overset{n}{\times}} (-\infty, x_i]\Big] \in \mathcal{Q}.$$

Since all sets of the form $\times_{i=1}^{n}(-\infty, x_i]$ generate the Borel field in $E^{(n)}$, we immediately obtain the conclusion by the definition of an n-dimensional random variable and by Proposition 1.

From measure theory we know that sums, products, quotients, and limits of sequences of measurable functions are measurable functions. The same then is true for random variables. The simplest kind of a random variable is the indicator, which is now defined.

Definition: If $A \in \mathcal{Q}$, then the indicator of A, I_A, is defined as a function over Ω that satisfies

$$I_A(\omega) = \begin{cases} 0 & \text{if } \omega \notin A \\ 1 & \text{if } \omega \in A. \end{cases}$$

One easily sees that I_A is indeed a random variable, since

$$[I_A \leq x] = \begin{cases} \phi & \text{if } x < 0 \\ A^c & \text{if } 0 \leq x < 1 \\ \Omega & \text{if } x \geq 1. \end{cases}$$

Definition: A random variable X is said to be discrete if there is a countable sequence of distinct numbers $\{a_n\}$ such that

$$1 = P(\cup_n [X = a_n]) = \sum_n P[X = a_n].$$

If X is discrete, as defined above, then X can be written in terms of indicators, namely,

$$X = \sum_n a_n I_{[X=a_n]},$$

except over an event of probability zero.

Definition: If X is a random variable, the sigma field induced by X, $\sigma\{X\}$, is defined by

$$\sigma\{X\} = \sigma\{[X \leq x], -\infty < x < \infty\}.$$

Proposition 4. If X is a random variable, and if $\mathfrak{B}$ denotes the sigma field of all Borel sets of real numbers, then

$$\sigma\{X\} = \{[X \in B], B \in \mathfrak{B}\}.$$

Proof: This is a corollary to Proposition 1.

We shall need one further definition for subsequent work.

Definition: The sigma field $\sigma\{X_\lambda, \lambda \in \Lambda\}$ induced (or determined or generated) by the set of random variables $\{X_\lambda, \lambda \in \Lambda\}$ is defined by

$$\sigma\{X_\lambda, \lambda \in \Lambda\} = \sigma\{\cup_{\lambda \in \Lambda} \sigma\{X_\lambda\}\}.$$

A theorem frequently used in statistics as well as in probability is the following.

Theorem 1. If X and Y are random variables, and if $\sigma\{Y\} \subset \sigma\{X\}$, then there exists a Borel measurable function f over $(-\infty, +\infty)$ such that $Y = f(X)$. Conversely, if X and Y are random variables, and if $Y = f(X)$, where f is a Borel measurable function over $(-\infty, +\infty)$, then $\sigma\{Y\} \subset \sigma\{X\}$.

Proof: Let us denote $B_{jm} = [j/2^m \leq Y < (j+1)/2^m]$ for $m = 1, 2, \cdots$, $j = 0, \pm 1, \pm 2, \cdots$. By definition of $\sigma\{Y\}$, $B_{jm} \in \sigma\{Y\}$. By hypothesis, $\sigma\{Y\} \subset \sigma\{X\}$, and hence $B_{jm} \in \sigma\{X\}$. Thus by Proposition 4 there exists a Borel set of real numbers C_{jm} such that $B_{jm} = [X \in C_{jm}]$. Now for fixed m, $\{B_{jm}\}$ are disjoint in j. However, the C_{jm} are not necessarily disjoint in j. So define

$$D_{jm} = C_{jm} \cap_{i \neq j} C_{im}^c.$$

Note that, since $B_{2j,m+1} \cup B_{2j+1,m+1} = B_{j,m}$, then the C_{jm} and therefore the D_{jm} can be selected so that

$$C_{2j,m+1} \cup C_{2j+1,m+1} = C_{j,m} \quad \text{and} \quad D_{2j,m+1} \cup D_{2j+1,m+1} = D_{j,m}.$$

(This follows from the fact that

$$[X \in A] \cup [X \in B] = [X \in A \cup B].$$

In the case of the $D_{j,m}$ we simply replace $D_{2j,m+1}$ and $D_{2j+1,m+1}$ by their intersections with D_{jm}.) Hence for every m,

$$D = \bigcup_{j=-\infty}^{\infty} D_{j1} = \bigcup_{j=-\infty}^{\infty} D_{jm},$$

and the D_{jm} are disjoint in j. We next show that $B_{jm} = [X \in D_{jm}]$. By the definition of D_{jm} given above, we have

$$[X \in D_{jm}] = X^{-1}(D_{jm}) = X^{-1}(C_{jm}) \cap_{i \neq j} X^{-1}(C_{im}^c)$$

$$= X^{-1}(C_{jm}) \left(\bigcup_{i \neq j} X^{-1}(C_{im}) \right)^c$$

$$= [X \in C_{jm}] \left(\bigcup_{i \neq j} [X \in C_{im}] \right)^c$$

$$= B_{jm} \left(\bigcup_{i \neq j} B_{im} \right)^c = B_{jm},$$

since for fixed m, $\{B_{km}\}$ are disjoint, and hence we have established that $B_{jm} = [X \in D_{jm}]$. Let

$$Y_m = \sum_{j=-\infty}^{\infty} \frac{j}{2^m} I_{B_{jm}}.$$

We observe that $|Y_m - Y| \leq 1/2^m$ over Ω for all m. Hence $Y_m \to Y$ everywhere over Ω as $m \to \infty$. Let us denote

$$f_m(x) = \sum_{j=-\infty}^{\infty} \frac{j}{2^m} I_{D_{jm}}(x).$$

Clearly $f_m(X) = Y_m$, so $|f_m(X) - Y| \leq 1/2^m$ over Ω. Due to the fact that

$$D_{jm} = D_{2j,m+1} \cup D_{2j+1,m+1},$$

we obtain that $\{f_m\}$ is uniformly mutually convergent. Hence there is a Borel-measurable function f such that $|f_m(x) - f(x)| \leq 1/2^m$ for all m and all $x \in (-\infty, +\infty)$. Finally we note that

$$|f(X) - Y| \leq |f(X) - f_m(X)| + |f_m(X) - Y| \leq 2/2^m \to 0$$

as $m \to \infty$, which proves that $Y = f(X)$ *everywhere* over Ω. Conversely, if $Y = f(X)$, if f is Borel measurable, and if $\mathcal{B}$ denotes the set of all Borel subsets of $(-\infty, +\infty)$, then by Proposition 4,

$$\sigma\{Y\} = Y^{-1}(\mathcal{B}) = f(X)^{-1}(\mathcal{B}) = X^{-1}(f^{-1}(\mathcal{B})).$$

But $f^{-1}(\mathcal{B}) \subset \mathcal{B}$ since f is Borel measurable. Hence

$$X^{-1}(f^{-1}(\mathcal{B})) \subset X^{-1}(\mathcal{B}) = \sigma\{X\},$$

which establishes the converse.

EXERCISES

1. Let $X_1, \cdots, X_n$ be n random variables. Prove that

$$\max\{X_i, 1 \leq i \leq n\} \quad \text{and} \quad \min\{X_i, 1 \leq i \leq n\}$$

are random variables.

2. Prove: if $A \in \mathcal{A}, B \in \mathcal{A}$, then

    (a) $I_A I_B = \min\{I_A, I_B\} = I_{AB}$,
    (b) $I_{A \cup B} = \max\{I_A, I_B\} = I_A + I_B - I_{AB}$ and
    (c) $I_{A \cup B} = I_A + I_B$ if and only if A and B are disjoint.

3. Prove: if $\{X_n\}$ is a sequence of random variables, and if

$$\mathcal{B}_n = \sigma\{X_k, k \geq n\},$$

then $\mathcal{B}_n \supset \mathcal{B}_{n+1}$ for all n.

4. Prove: if $\mathcal{C}$ is a collection of subsets of Ω, and if ω' and ω'' are two fixed points in Ω such that for every $C \in \mathcal{C}$, either $\{\omega', \omega''\} \subset C$ or $\{\omega', \omega''\} \subset \Omega \backslash C$, then the same property holds for every $C \in \sigma\{\mathcal{C}\}$.

5. If $\mathbf{X} = (X_1, \cdots, X_n)$ is an n-dimensional random variable, then for $1 \leq i_1 < i_2 < \cdots < i_k \leq n$, the k-tuple $\mathbf{Y} = (X_{i_1}, \cdots, X_{i_k})$ is a k-dimensional random variable.

CHAPTER 2

Probability Distributions

2.1. Univariate Distribution Functions

In the first chapter we introduced the concepts of probability space and random variable. In this chapter we deal with distribution functions, their moments and their transforms. This section is devoted to an introduction to univariate distribution functions.

Definition: If X is a random variable, its distribution function F_X is defined by $F_X(x) = P[X \leq x]$ for all $x \in (-\infty, +\infty)$.

Note that different random variables can have the same distribution functions. For example, if Ω contains only two elementary events, that is, $\Omega = \{H, T\}$, if $P(H) = P(T) = \frac{1}{2}$, and if X and Y are random variables defined by $X(H) = 1$, $X(T) = 0$, $Y(H) = 0$, and $Y(T) = 1$, then it is easy to verify that

$$F_X(x) = F_Y(x) = \begin{cases} 0 & \text{if } x < 0 \\ \frac{1}{2} & \text{if } 0 \leq x < 1 \\ 1 & \text{if } x \geq 1. \end{cases}$$

Theorem 1. If X is a random variable, then its distribution function F_X has the following properties:

(a) F_X is nondecreasing; that is, if $-\infty < x' < x'' < \infty$, then

$$F_X(x') \leq F_X(x''),$$

(b) $\lim_{x \to \infty} F_X(x) = 1$ and $\lim_{x \to -\infty} F_X(x) = 0$, and
(c) F_X is continuous from the right, that is,

$$\lim_{h \downarrow 0} F_X(x + h) = F_X(x) \qquad \text{for all } x.$$

Proof: (a) It can be easily verified that $[X \leq x'] \subset [X \leq x'']$. Applying the monotonicity property of probability (Theorem 3 of Section 1.2) we obtain (a). (b) Since $X(\omega)$ is real for every $\omega \in \Omega$, then $[X \leq -n] \to \phi$ as $n \to \infty$, which implies, by taking probabilities, that $F_X(x) \to 0$ as $x \to -\infty$. Similarly, $[X \leq n] \to \Omega$ as $n \to \infty$ yields $F_X(x) \to 1$ as $x \to \infty$. (c) Let $h > 0$. Then

$$F_X(x + h) - F_X(x) = P[x < X \leq x + h].$$

But $[x < X \leq x + h] \to \phi$ as $h \downarrow 0$, which concludes the proof.

It will be recalled from real analysis or advanced calculus that a function of bounded variation, which is what $F_X(x)$ is, has at most a countable number of discontinuities. Further, all discontinuities of such functions are jumps. This fact will be used repeatedly in this text.

Theorem 1 characterizes probability distribution functions in a sense that is made precise by the following theorem.

Theorem 2. If F is a function defined over $(-\infty, +\infty)$ which satisfies (a), (b), and (c) of Theorem 1, then there exists a probability space $(\Omega, \mathcal{C}, P)$ and a random variable X defined over Ω such that $F_X(x) = F(x)$ for all real x.

Proof: Select $\Omega = (-\infty, +\infty)$, let $\mathcal{C}$ denote the sigma field of Borel subsets of $(-\infty, +\infty)$, and let P be the Lebesgue-Stieltjes measure determined by F over $\mathcal{C}$. Define X by $X(\omega) = \omega$ for all $\omega \in (-\infty, +\infty)$. It is easily verified that $P[X \leq x] = F(x)$, thus giving us the conclusion of the theorem.

Definition: A distribution function F is said to be absolutely continuous if there exists a Borel measurable function f over $(-\infty, +\infty)$ such that

$$F(x) = \int_{-\infty}^{x} f(t) \, dt$$

for all real x. The function f is called a density of F.

Actually, f is almost everywhere (with respect to Lebesgue measure) uniquely determined by the Radon-Nikodym theorem, and indeed it is the Radon-Nikodym derivative of the Lebesgue-Stieltjes measure determined by F with respect to Lebesgue measure. It should be remarked that if $g \in L^1(-\infty, +\infty)$, if $g \geq 0$ almost everywhere, and if

$$\int_{-\infty}^{\infty} g(x) \, dx = 1,$$

then g determines an absolutely continuous distribution function G,

$$G(x) = \int_{-\infty}^{x} g(t) \, dt,$$

and g is its density. Below are listed some of the most frequently encountered densities of absolutely continuous distribution functions.

(a) *The normal or Gaussian distribution.* This is the most important of all such distribution functions. Its density is given by

$$f(x) = (2\pi\sigma^2)^{-1/2} \exp \{-(x - \mu)^2/2\sigma^2\}$$

for $-\infty < x < \infty$, where μ and σ^2 are fixed constants, $-\infty < \mu < \infty$ and $\sigma^2 > 0$. This distribution will be referred to as $\mathfrak{N}(\mu, \sigma^2)$.

(b) *The uniform distribution over* $[a, b]$:

$$f(x) = \begin{cases} 1/(b - a) & \text{if } a < x < b \\ 0 & \text{if } x \leq a \text{ or } x \geq b, \end{cases}$$

where $a < b$ are fixed constants.

(c) The gamma distribution:

$$f(x) = \begin{cases} (1/\Gamma(\alpha + 1)\beta^{\alpha+1})x^{\alpha} \exp{(-x/\beta)} & \text{if} \quad x > 0 \\ 0 & \text{if} \quad x \leq 0, \end{cases}$$

where $\alpha > -1$ and $\beta > 0$ are two fixed constants.

(d) The beta distribution:

$$f(x) = \begin{cases} x^{p-1}(1 - x)^{q-1}/B(p, q) & \text{if} \quad 0 < x < 1 \\ 0 & \text{if} \quad x \leq 0 \quad \text{or} \quad x \geq 1, \end{cases}$$

where $p > 0, q > 0$ are two fixed real numbers and

$$B(p, q) = \int_0^1 x^{p-1}(1 - x)^{q-1} \, dx$$

is the beta function.

(e) The negative exponential distribution:

$$f(x) = \begin{cases} \alpha \exp{[-\alpha(x - \beta)]} & \text{if} \quad x > \beta \\ 0 & \text{if} \quad x \leq \beta, \end{cases}$$

where $\alpha > 0$ and $-\infty < \beta < \infty$ are two fixed constants.

(f) The Cauchy distribution:

$$f(x) = 1/\pi(1 + x^2), \qquad -\infty < x < \infty.$$

Definition: A distribution function F is said to be discrete if there exists a countable sequence $\{x_n\}$ of real numbers and a corresponding sequence $\{p_n\}$ of positive numbers such that

$$\sum p_n = 1 \quad \text{and} \quad F(x) = \sum \{p_n \mid x_n \leq x\}.$$

Lemma 1. If X is a random variable whose distribution function has a discontinuity (jump) of size γ, $0 < \gamma < 1$, at x_0, that is, if $F_X(x_0) - F_X(x_0 - 0) = \gamma$, then $P[X = x_0] = \gamma$.

Proof: Since

$$[x_0 - 1/n < X \leq x_0] \to [X = x_0] \qquad \text{as } n \to \infty,$$

since

$$P[x_0 - 1/n < X \leq x_0] = F_X(x_0) - F(x_0 - 1/n),$$

and since

$$F_X(x_0 - 1/n) \to F_X(x_0 - 0) \qquad \text{as } n \to \infty,$$

we obtain

$$P[X = x_0] = F_X(x_0) - F_X(x_0 - 0).$$

Proposition 1. The distribution function F_X of a random variable X is discrete if and only if X is discrete.

Proof: This follows immediately from the lemma and the definitions of discreteness of F_X and of X.

Some of the more important discrete distribution functions are the following:

(a) *The binomial distribution*

$$F(x) = \begin{cases} 0 & \text{if } x < 0 \\ \sum_{0 \leq k \leq x} \binom{n}{k} p^k (1-p)^{n-k} & \text{if } 0 \leq x \leq n, \\ 1 & \text{if } x \geq n \end{cases}$$

where $0 < p < 1$ and n is a positive integer. This distribution will be denoted by $B(n, p)$.

(b) *The Poisson distribution*

$$F(x) = \begin{cases} 0 & \text{if } x < 0 \\ \sum_{0 \leq n \leq x} e^{-\lambda} \lambda^n / n! & 0 \leq x < \infty \end{cases}$$

where $\lambda > 0$ is a constant. The distribution function will be denoted by $\mathcal{P}(\lambda)$.

(c) **The geometric distribution**

$$F(x) = \begin{cases} 0 & \text{if } x < 0 \\ \sum_{0 \le n \le x} p(1-p)^n & \text{if } x \ge 0 \end{cases}$$

where $0 < p < 1$.

(d) **The uniform distribution** over $\{1, 2, \cdots, n\}$

$$F(x) = \begin{cases} 0 & \text{if } x < 0 \\ [x]/n & \text{if } 0 \le x \le n, \\ 1 & \text{if } n < x, \end{cases}$$

where $[x]$ denotes the largest integer equal to or less than x.

Definition: A distribution function F is said to be continuous singular if it is continuous and if there exists a Borel set of real numbers S of Lebesgue measure zero such that $\mu_F(S) = 1$, where μ_F is the Lebesgue-Stieltjes measure determined by F.

The most frequent example of a continuous singular distribution function is the Cantor distribution, or uniform distribution over the Cantor ternary set. Its construction is outlined as follows:

(1) Let

$$f_1(x) = \begin{cases} \frac{3}{2} & \text{if } x \in [0, \frac{1}{3}] \cup [\frac{2}{3}, 1] = E_1 \\ 0 & \text{if } x \notin E_1, \end{cases}$$

and define

$$F_1(x) = \int_{-\infty}^{x} f_1(t)\, dt.$$

(n) Let E_n denote the union of 2^n disjoint intervals of length $(\frac{1}{3})^n$ obtained by deleting the middle thirds from the 2^{n-1} disjoint intervals of length $(\frac{1}{3})^{n-1}$ in E_{n-1}. Define

$$f_n(x) = \begin{cases} (\frac{3}{2})^n & \text{if } x \in E_n \\ 0 & \text{if } x \notin E_n, \end{cases}$$

and define

$$F_n(x) = \int_{-\infty}^{x} f_n(t) \, dt.$$

It is now easy to show that the distribution functions $\{F_n\}$ are continuous and converge uniformly; more precisely, if $m < n$, then

$$| F_m(x) - F_n(x) | \leq (\tfrac{1}{2})^m \qquad \text{for all } x.$$

Hence there exists a continuous function F such that $F(x) = 0$ if $x \leq 0$, $F(x) = 1$ if $x \geq 1$, and such that $F_m(x) \to F(x)$ as $m \to \infty$ uniformly in x. Further, the Lebesgue-Stieltjes measure μ_F determined by F gives measure zero to the complement of the Cantor ternary set $C = \cap_{n=1}^{\infty} E_n$. Hence $\mu_F(C) = 1$ although the Lebesgue measure of C is easily shown to be zero.

Of course, a distribution function need not fall into any of the three pure categories listed above but can be a mixture of any two or all three; indeed, the only properties a distribution function need have are the three given in Theorem 1.

We conclude this section by considering the convolution of two distribution functions.

Definition: If F and G are two distribution functions, their convolution $F * G$ is a function defined by

$$F * G(x) = \int_{-\infty}^{\infty} F(x - y) \, dG(y), \qquad -\infty < x < \infty,$$

where the integral is a Lebesgue-Stieltjes integral.

Proposition 2. If F and G are distribution functions, then $F * G$ is a distribution function.

Proof: By Theorem 2 we need only verify requirements (a), (b), and (c) of Theorem 1.

(a) If $x' < x''$, then

$$F * G(x'') - F * G(x') = \int_{-\infty}^{\infty} \{F(x'' - y) - F(x' - y)\} \, dG(y).$$

Since F satisfies (a) in Theorem 1, the integrand in the above integral is nonnegative, and hence so is the integral.

(b) Both limits in this requirement are satisfied by applying the Lebesgue dominated convergence theorem.

(c) is also satisfied by means of the Lebesgue dominated convergence theorem. Thus $F * G$ is a distribution function.

EXERCISES

1. Let Ω denote the set of all outcomes when tossing an unbiased coin twice, and let X denote the number of heads observed. What are Ω, $\mathcal{C}$, P, $\{X(\omega), \omega \in \Omega\}$, and F_X?

2. **Definition:** A number x is called a point of increase of a distribution function F if $F(x + \epsilon) - F(x - \epsilon) > 0$ for all $\epsilon > 0$. Construct a discrete distribution function F such that every real number is a point of increase.

3. Prove: if

$$p_n \geq 0, \qquad \sum_{n=0}^{\infty} p_n = 1, \qquad q_n \geq 0, \qquad \sum_{n=0}^{\infty} q_n = 1,$$

if

$$F(x) = \sum \{p_n \mid 0 \leq n \leq x\},$$

and if

$$G(x) = \sum \{q_n \mid 0 \leq n \leq x\},$$

then

$$F * G(x) = \sum \{ \sum_{k=0}^{n} p_k q_{n-k} \mid 0 \leq n \leq x\}.$$

4. Prove: if F, G are distribution functions and if F is absolutely con-

tinuous with density f, then $F * G$ is absolutely continuous, and its density is

$$h(x) \ = \ \int_{-\infty}^{\infty} f(x - y) \, dG(y),$$

where the integral is a Lebesgue-Stieltjes integral.

5. Prove: If X and Y are random variables, if F_X is absolutely continuous, and if F_Y is discrete, then F_{X+Y} is absolutely continuous.

6. If X is a random variable with a continuous distribution function F, find the distribution function of the random variable $F(X)$.

7. Let X be a random variable whose distribution function is uniform over $[0, 1]$. Let G be any distribution function. Show that there exists a Borel-measurable function g such that G is the distribution function of $g(X)$.

8. In the construction given of the Cantor distribution function, prove that E_n is the closure of the set of x in $[0, 1]$ for which the first n digits in the triadic expansions are 0 and 2.

2.2. Multivariate Distribution Functions

In this section we extend the notion of distribution function of one random variable to that of the joint distribution function of any finite number of random variables. Our aim in this section is to characterize these joint distribution functions.

Definition: Let $X_1, X_2, \cdots, X_n$ be random variables, $n \geq 1$. The joint distribution function of $X_1, \cdots, X_n$, or the distribution function of the random vector $\mathbf{X} = (X_1, \cdots, X_n)$, is defined to be

$$F_{\mathbf{X}}(x_1, \cdots, x_n) \ = \ P(\bigcap_{i=1}^{n} [X_i \leq x_i]),$$

where $-\infty < x_i < \infty, 1 \leq i \leq n$.

Lemma 1. If $\mathbf{X} = (X_1, \cdots, X_n)$ is a vector random variable, then, for $n \geq 2$,

$$\lim_{x_n \to \infty} F_{X_1,\cdots,X_n}(x_1, \cdots, x_n) \ = \ F_{X_1,\cdots,X_{n-1}}(x_1, \cdots, x_{n-1}).$$

Proof: The proof of this lemma is the same as the proof of (b) in Theorem 1 in Section 2.1.

The distribution function $F_{X_1,\cdots,X_{n-1}}$ will be referred to as a *marginal* or *marginal distribution* of $F_{X_1,\cdots,X_n}$. In general the joint distribution of any subset of the random variables $X_1, \cdots, X_n$ will be referred to as a marginal or marginal distribution function of $F_{X_1,\cdots,X_n}$.

For what follows, some vector notation is needed. Let $a_i < b_i$, $1 \leq i \leq n$, be n pairs of real numbers, where possibly also $b_i = +\infty$ and $a_i = -\infty$. Let us denote $\mathbf{a} = (a_1, a_2, \cdots, a_n)$, $\mathbf{b} = (b_1, b_2, \cdots, b_n)$, and

$$(\mathbf{a}, \mathbf{b}] = \{\mathbf{x} = (x_1, \cdots, x_n) \mid a_i < x_i \leq b_i, 1 \leq i \leq n\}.$$

We shall refer to $(\mathbf{a}, \mathbf{b}]$ as a *cell* in $E^{(n)}$. Let $\Delta_{k,n}$ denote the set of $\binom{n}{k}$ n-tuples $(z_1, \cdots, z_n)$ where each z_i is a_i or b_i and such that exactly k of the z_i are a_i. Then $\Delta \doteq \cup_{k=0}^{n}\Delta_{k,n}$ is the set of the 2^n vertices of the cell $(\mathbf{a}, \mathbf{b}]$ in $E^{(n)}$, and an arbitrary vertex will be denoted by δ.

Theorem 1. If $\mathbf{X} = (X_1, \cdots, X_n)$ is an n-dimensional random variable, then

(a) $\lim_{\min x_i \to \infty} F_X(x_1, \cdots, x_n) = 1$,
(b) for each i, $1 \leq i \leq n$, $\lim_{x_i \to -\infty} F_X(x_1, \cdots, x_n) = 0$,
(c) $F_X(x_1, \cdots, x_n)$ is continuous from above in each argument, and
(d) for every cell $(\mathbf{a}, \mathbf{b}]$ in $E^{(n)}$,

$$\mu_F(\mathbf{a}, \mathbf{b}] \doteq \sum_{k=0}^{n} (-1)^k \sum_{\delta \in \Delta_{k,n}} F_X(\delta) \geq 0,$$

where μ_F is defined by this equality.

Proof: The proofs of (a), (b), and (c) are the same as the proofs of (b) and (c) in Theorem 1 in Section 2.1 and will not be repeated here. In order to prove (d), we readily see by Lemma 1 that we need only prove that for any other m random variables $\mathbf{Y} = (Y_1, \cdots, Y_m)$, and all m, that

$$\sum_{k=0}^{n} (-1)^k \sum_{\delta \in \Delta_{k,n}} F_{X,Y}(\delta, \mathbf{c}) \geq 0, \tag{1}$$

where $\mathbf{c} = (c_1, \cdots, c_m)$. This is so because, by Lemma 1, we can take the limit of (1) as min $\{c_1, \cdots, c_m\} \to \infty$ and obtain (d). In order to prove (1) it is sufficient to prove that the sum on the left is equal to

$$P((\bigcap_{i=1}^{n} [a_i < X_i \leq b_i])(\bigcap_{j=1}^{m} [Y_j \leq c_j])).$$

We do this by induction on n. It is easy to verify that this is true for $n = 1$ and all m. We now assume it is true for some arbitrary n and all m and show that this implies the equality for $n + 1$ and all m. Indeed

$$\sum_{k=0}^{n+1} (-1)^k \sum_{\delta \in \Delta_{k,n+1}} F_{X,Y}(\delta, \mathbf{c})$$

$$= \sum_{k=1}^{n} (-1)^k \sum_{\delta \in \Delta_{k,n+1}} F_{X,Y}(\delta, \mathbf{c})$$

$$+ (-1)^{n+1} F_{X,Y}(a_1, \cdots, a_{n+1}, \mathbf{c})$$

$$+ F_{X,Y}(b_1, \cdots, b_{n+1}, \mathbf{c})$$

$$= \sum_{k=1}^{n} (-1)^k \sum_{\delta \in \Delta_{k,n}} F_{X,Y}(\delta, b_{n+1}, \mathbf{c})$$

$$+ \sum_{k=1}^{n} (-1)^k \sum_{\delta \in \Delta_{k-1,n}} F_{X,Y}(\delta, a_{n+1}, \mathbf{c})$$

$$+ (-1)^{n+1} F_{X,Y}(a_1, \cdots, a_{n+1}, \mathbf{c})$$

$$+ F_{X,Y}(b_1, \cdots, b_{n+1}, \mathbf{c}),$$

and, combining the first and fourth expressions and then the second and third expressions, the above

$$= \sum_{k=0}^{n} (-1)^k \sum_{\delta \in \Delta_{k,n}} F_{X,Y}(\delta, b_{n+1}, \mathbf{c})$$

$$- \sum_{k=0}^{n} (-1)^k \sum_{\delta \in \Delta_{k,n}} F_{X,Y}(\delta, a_{n+1}, \mathbf{c}),$$

and using induction hypothesis

$$= P((\bigcap_{i=1}^{n} [a_i < X_i \le b_i])[X_{n+1} \le b_{n+1}] \bigcap_{j=1}^{m} [Y_j \le c_j])$$

$$- P((\bigcap_{i=1}^{n} [a_i < X_i \le b_i])[X_{n+1} \le a_{n+1}] \bigcap_{j=1}^{m} [Y_j \le c_j])$$

$$= P((\bigcap_{i=1}^{n+1} [a_i < X_i \le b_i])(\bigcap_{j=1}^{m} [Y_j \le c_j])) \ge 0,$$

which concludes the induction and the proof.

It should be noticed that (d) easily implies that F_X is nondecreasing in each argument. However, a function which is nondecreasing in each argument does not necessarily satisfy (d). As an example, let $f(x, y)$ be defined by

$$f(x, y) = \begin{cases} 1 & \text{if } \max\{x, y\} \ge 0 \quad \text{or} \quad x^2 + y^2 \le 1 \\ 0 & \text{otherwise} \end{cases}$$

Clearly this function is nondecreasing in each variable. If we take $a = (-\frac{3}{4}, -\frac{3}{4})$ and $b = (-\frac{1}{4}, -\frac{1}{4})$, then $\mu_F(a, b]$ as defined in (d) is easily seen to be equal to -1, and hence does not satisfy (d).

We now show that requirements (a)–(d) in Theorem 1 characterize joint distribution functions of n-dimensional random vectors.

Theorem 2. Let F be a real-valued function defined over $E^{(n)}$ which satisfies:

(a) $F(x_1, \cdots, x_n) \to 1$ as $\min\{x_1, \cdots, x_n\} \to \infty$,

(b) for each i, $1 \le i \le n$, $F(x_1, \cdots, x_n) \to 0$ as $x_i \to -\infty$,

(c) $F(x_1, \cdots, x_n)$ is continuous from the right in each argument, and

(d) $\sum_{k=0}^{n} (-1)^k \sum_{\delta \in \Delta_{k,n}} F(\delta) \ge 0$ for every cell $(a, b]$ in $E^{(n)}$.

Then there exist a probability space $(\Omega, \mathcal{C}, P)$ and n random variables $X_1, \cdots, X_n$ defined over Ω such that $F_{X_1, \cdots, X_n}(x_1, \cdots, x_n) = F(x_1, \cdots, x_n)$.

Proof: Let us denote, for every cell $(a, b]$ in $E^{(n)}$,

$$\mu_F(a, b] = \sum_{k=0}^{n} (-1)^k \sum_{\delta \in \Delta_{k,n}} F(\delta).$$

Let $a_1 < c_1 < b_1$, and denote

$$\mathbf{a}^* = (c_1, a_2, \cdots, a_n) \qquad \text{and} \qquad \mathbf{b}^* = (c_1, b_2, \cdots, b_n).$$

Then

$$\mu_F(\mathbf{a}, \mathbf{b}^*] + \mu_F(\mathbf{a}^*, \mathbf{b}] = \mu_F(\mathbf{a}, \mathbf{b}],$$

since the terms in the sums determining $\mu_F(\mathbf{a}, \mathbf{b}^*]$ and $\mu_F(\mathbf{a}^*, \mathbf{b}]$ which involve c_1 and the same a_i and b_i are of opposite sign, since -1 is raised to a power one larger in the second than in the first, and hence all terms involving c_1 cancel. Consequently μ_F is an interval function and determines a Lebesgue-Stieltjes measure P over the Borel sets in $E^{(n)}$. By (a), (b), and (d), $P(E^{(n)}) = 1$. Now set $\Omega = E^{(n)}$, let $\mathcal{a}$ be the sigma field of Borel sets in $E^{(n)}$, and let P be as defined. For every

$$\omega = (x_1, \cdots, x_n) \in \Omega,$$

define $X_i(\omega) = x_i$, $1 \leq i \leq n$. Then by (c) and (d),

$$P(\bigcap_{i=1}^{n} [X_i \leq x_i]) = P(\mathbf{a}, \mathbf{b}] = F(x_1, \cdots, x_n),$$

where $a_i = -\infty$ and $b_i = x_i$, $1 \leq i \leq n$, which concludes the proof.

It was noted earlier that the distribution function of one random variable is discontinuous at most at a countable number of points. At a later state it will be necessary for us to know the nature of the set of points at which a multivariate distribution is not continuous. The answer to this is given in the following theorem.

Theorem 3. Let F be a joint distribution function of n random variables. Then F is continuous everywhere except over the union of a countable set of hyperplanes of the form $x_j = c$, $1 \leq j \leq n$.

Proof: Let $\mathbf{X} = (X_1, \cdots, X_n)$ be the n random variables of which F is the joint distribution function. Let $F_i(x) = P[X_i \leq x]$. As was previously pointed out, F_i has at most a countable number of discontinuities. Denote the values of x at which the discontinuities of F_i occur $\{c_1^{(i)}, c_2^{(i)}, \cdots\}$. Let

$$D = \bigcup_{i=1}^{n} \bigcup_{m} \{(x_1, \cdots, x_n) \mid x_i = c_m^{(i)}\}.$$

We now prove that F is continuous over $E^{(n)} \backslash D$. Let

$$\mathbf{x} = (x_1, \cdots, x_n) \in E^{(n)} \backslash D.$$

We first observe that F_j is continuous at x_j, $1 \leq j \leq n$. Now for $\mathbf{h} = (h_1, \cdots, h_n)$, we have

$$| F(\mathbf{x} + \mathbf{h}) - F(\mathbf{x}) |$$

$$\leq F(x_1 + | h_1 |, \cdots, x_n + | h_n |) - F(x_1 - | h_1 |, \cdots, x_n - | h_n |)$$

$$= P(\bigcap_{j=1}^{n} [X_j \leq x_j + | h_j |]) - P(\bigcap_{j=1}^{n} [X_j \leq x_j - | h_j |])$$

$$= P((\bigcap_{i=1}^{n} [X_i \leq x_i + | h_i |]) (\bigcap_{j=1}^{n} [X_j \leq x_j - | h_j |])^c)$$

$$= P((\bigcap_{i=1}^{n} [X_i \leq x_i + | h_i |]) (\bigcup_{j=1}^{n} [X_j > x_j - | h_j |]))$$

$$= P(\bigcup_{j=1}^{n} [X_j > x_j - | h_j |] \cap [X_i \leq x_i + | h_i |]))$$

$$\leq P(\bigcup_{j=1}^{n} [x_j - | h_j | < X_j \leq x_j + | h_j |])$$

$$\leq \sum_{j=1}^{n} (F_j(x_j + | h_j |) - F_j(x_j - | h_j |)).$$

Since F_i is continuous at x_i, $1 \leq i \leq n$, the conclusion of the theorem follows from the last inequality by taking limits as $\mathbf{h} \to \mathbf{0}$, where $\mathbf{0} = (0, 0, \cdots, 0)$.

As with the univariate case, so may we consider Lebesgue properties in the multivariate case. A joint distribution function F of n random variables is said to be *absolutely continuous* if there exists a Borel measurable function f defined over $E^{(n)}$ such that

$$F(x_1, \cdots, x_n) = \int_{-\infty}^{x_1} \cdots \int_{-\infty}^{x_n} f(t_1, \cdots, t_n) \, dt_1 \cdots dt_n$$

for all $(x_1, \cdots, x_n) \in E^{(n)}$. It is said to be discrete if there exists a countable subset $S \subset E^{(n)}$ such that $\mu_F(S) = 1$, where μ_F is the Lebesgue-Stieltjes measure over $E^{(n)}$ determined by F. Last, F is said to be continuous singular if F is continuous and if there exists a Borel set S in $E^{(n)}$ of Lebesgue measure zero such that $\mu_F(S) = 1$.

EXERCISES

1. Let X, Y be random variables such that

$$F_{X,Y}(x, y) = \int_{-\infty}^{x} \int_{-\infty}^{y} f(u, v) \, du \, dv,$$

where

$$f(u, v) = \begin{cases} 1 & \text{if} \quad 0 \leq v \leq 2u, \quad 0 \leq u \leq 1 \\ 0 & \text{otherwise.} \end{cases}$$

Find F_X and F_Y.

2. Let F be a distribution function of one random variable, and let $G(x_1, \cdots, x_n) = \prod_{j=1}^{n} F(x_j)$. Verify that G is a distribution function of an n-dimensional random vector.

3. Use the conditions given in the hypothesis of Theorem 2 to prove that F is nondecreasing in each argument.

4. Prove: A multivariate distribution function is discrete if and only if every one-dimensional marginal distribution function is discrete.

2.3. Distribution of a Set of Infinitely Many Random Variables

In the previous two sections we have dealt with distribution functions, or distributions, of one and of a finite number of random variables. In this section we extend our consideration to distributions of any set of random variables. We shall define what a distribution of any set of random variables is and shall then characterize such distributions. This characterization is usually referred to as the Kolmogorov-Daniell extension theorem.

Definition: If $\mathfrak{X} = \{X_\lambda, \lambda \in \Lambda\}$ is any set of random variables, then the distribution of $\mathfrak{X}$ is defined to be the set of all joint distribution functions of all finite subsets of $\mathfrak{X}$.

If Λ is finite in the above definition, then given the joint distribution function of all the random variables we can by Lemma 1 in Section 2.2 determine the marginals and therefore the distribution in the above sense.

Theorem 1. Let $\mathfrak{X} = \{X_\lambda, \lambda \in \Lambda\}$ be a set of random variables, and let $F_{\lambda_1,\cdots,\lambda_n}(x_1, \cdots, x_n)$ denote the joint distribution function of $X_{\lambda_1}, \cdots, X_{\lambda_n}$. Then for every finite subset $\{\lambda_1, \cdots, \lambda_n\}$ of Λ, conditions (a), (b), (c), (d) of Theorem 1 in Section 2.2 are satisfied, and, in addition,

(e) $\qquad \lim_{x_1 \to \infty} F_{\lambda_1,\cdots,\lambda_n}(x_1, \cdots, x_n) = F_{\lambda_2,\cdots,\lambda_n}(x_2, \cdots, x_n)$

and

(f) $\qquad F_{\lambda_{i_1},\cdots,\lambda_{i_n}}(x_{i_1}, \cdots, x_{i_n}) = F_{\lambda_1,\cdots,\lambda_n}(x_1, \cdots, x_n)$

for all permutations $(i_1, \cdots, i_n)$ of $(1, 2, \cdots, n)$.

Proof: Properties (a)–(e) have already been proved in Lemma 1 and Theorem 1 of Section 2.2, and property (f) is a trivial consequence of the definition of joint distribution function.

Conditions (a)–(f) of Theorem 1 characterize the distribution of an infinite set of random variables in a certain sense which is made precise by the following theorem.

Theorem 2 (Daniell, Kolmogorov). If Λ is a set, and if

$$\{F_{\lambda_1,\cdots,\lambda_n}(x_1, \cdots, x_n), \quad \text{all finite subsets } \{\lambda_1, \cdots, \lambda_n\} \subset \Lambda\}$$

is a set of functions which satisfy conditions (a)–(f) in Theorem 1, then there exists a probability space $(\Omega, \mathcal{C}, P)$ and a set of random variables $\{X_\lambda, \lambda \in \Lambda\}$ over Ω such that $F_{\lambda_1,\cdots,\lambda_n}$ is the joint distribution function of $\{X_{\lambda_1}, \cdots, X_{\lambda_n}\}$ for every finite subset $\{\lambda_1, \cdots, \lambda_n\}$ of Λ.

Proof: For every $\lambda \in \Lambda$, let $R_\lambda = (-\infty, +\infty)$. Since

$$F_{\lambda_1,\cdots,\lambda_n}(x_1, \cdots, x_n) \text{ satisfies (a)–(d)},$$

then by Theorem 2 in Section 2.2 it determines a probability measure $P_{\lambda_1,\cdots,\lambda_n}$ over all Borel subsets of n-dimensional Euclidean space $\times_{i=1}^{n} R_{\lambda_i}$. If B is a Borel set in $\times_{i=1}^{n-1} R_{\lambda_i}$, then by (e) (the consistency condition)

$$P_{\lambda_1,\cdots,\lambda_{n-1}}(B) = P_{\lambda_1,\cdots,\lambda_n}(B \times R_{\lambda_n}).$$

Denote $\Omega = \times_{\lambda \in \Lambda} R_\lambda$, and let $\mathcal{C}_0$ denote the collection of those subsets of Ω of the form

$$A = B \times (\times \{R_\lambda \mid \lambda \in \Lambda \setminus \{\lambda_1, \cdots, \lambda_n\}\}),$$

where B is a Borel set in $\times_{i=1}^{n} R_{\lambda_i}$. It is clear that $\mathcal{C}_0$ is a field of subsets; that is, it is closed under finite unions and complementation. For every such A in $\mathbb{Q}_0$ we define

$$P_0(A) = P_{\lambda_1,\cdots,\lambda_n}(B).$$

Clearly P_0 is finitely additive over $\mathbb{Q}_0$, it is nonnegative, $P_0(\phi) = 0$, and $P_0(\Omega) = 1$. At this point we wish to apply the following theorem from measure theory: if M is a set, if $\mathfrak{M}$ is a field of subsets of M, and if μ is a sigma-finite measure over $\mathfrak{M}$, then there exists a unique extension of μ to a measure μ^* over $\mathfrak{M}^*$, where $\mathfrak{M}^*$ is the sigma field generated by $\mathfrak{M}$. We recall that μ being a measure over $\mathfrak{M}$ means that if $\{A_n\}$ is any denumerable sequence of disjoint sets in $\mathfrak{M}$, and *if* $\cup A_n \in \mathfrak{M}$, then $\mu(\cup_n A_n) = \sum_n \mu(A_n)$. A property equivalent to this is: if $A_n \in \mathfrak{M}$, if $A_n \supset A_{n+1}$ for all n, and if $A_n \to \phi$ as $n \to \infty$, then $\mu(A_n) \to 0$. Now let us return to the proof of the theorem. Let $A_n \in \mathbb{Q}_0$, $A_n \supset A_{n+1}$ for all n, and assume $A_n \to \phi$ as $n \to \infty$; we wish to prove that $P_0(A_n) \to 0$ as $n \to \infty$. Note: if only we were certain that the A_n were all cylinder sets depending on a common finite set of coordinates, then there would be nothing to prove, since $P_{\lambda_1,\cdots,\lambda_n}$ is a measure for every finite subset $\{\lambda_1, \cdots, \lambda_n\}$ of Λ. Let us therefore suppose, to the contrary, that there exists a $K > 0$ such that $\lim_{n\to\infty} P_0(A_n) \geq K$. We shall show a contradiction by proving that $\cap_{n=1}^{\infty} A_n \neq \phi$. Let $\{k_n\}$ be an increasing sequence of positive integers such that for some sequence of distinct elements $\{\lambda_n\}$ in Λ we may write

$$A_n = B_n \times (\times \{R_\lambda \mid \lambda \in \Lambda \backslash \{\lambda_1, \cdots, \lambda_{k_n}\}\}),$$

where B_n is a Borel set in $\times_{i=1}^{k_n} R_{\lambda_i}$. Now

$$P_0(A_n) = P_{\lambda_1,\cdots,\lambda_{k_n}}(B_n)$$

for all n. For every n there exists a *closed, bounded* set C_n such that

$$C_n \subset B_n \subset \times_{i=1}^{k_n} R_{\lambda_i} \quad \text{and} \quad P_{\lambda_1,\cdots,\lambda_{k_n}}(B_n \backslash C_n) < \epsilon/2^n,$$

where $0 < \epsilon < K$. Now let

$$D_n = C_n \times (\times \{R_\lambda \mid \lambda \in \Lambda \backslash \{\lambda_1, \cdots, \lambda_{k_n}\}\}).$$

Then it follows that $P_0(A_n \backslash D_n) < \epsilon/2^n$. Let $E_n = \cap_{j=1}^n D_j$. Then, since $A_n \supset A_{n+1}$ for all n,

$$P_0(A_n \backslash E_n) = P_0(A_n \cap (\cup_{j=1}^n D_j^c))$$

$$= P_0(\cup_{j=1}^n (A_n \backslash D_j)) \leq P_0(\cup_{j=1}^n (A_j \backslash D_j))$$

$$\leq \sum_{j=1}^n P_0(A_j \backslash D_j) < \epsilon.$$

Since $E_n \subset D_n \subset A_n$, then this last inequality yields $P_0(E_n) \geq P_0(A_n) - \epsilon \geq K - \epsilon > 0$. We may suppose without loss of generality that $k_n = n$ by simply inserting between A_j and A_{j+1} as many sets A_j as needed. Each E_n is nonempty since $P_0(E_n) > 0$. Consequently, for each n, there exists an $\mathbf{x}^{(n)} \in E_n \subset D_k$ for $1 \leq k \leq n$. Denote the λth coordinate of $\mathbf{x}^{(n)}$ by $x_\lambda^{(n)}$. For each k,

$$(x_{\lambda_1}^{(n)}, \cdots, x_{\lambda_k}^{(n)}) \in C_k$$

for all $n \geq k$. Hence, as a sequence in n, $\{x_{\lambda_1}^{(n)}\}$ is in a closed bounded set C_1 in R_{λ_1} and hence contains a subsequence $\{x_{\lambda_1}^{(1,n)}\}$ which converges to $w_1 \in C_1$. Since $\{(x_{\lambda_1}^{(1,n)}, x_{\lambda_2}^{(1,n)})\}$ is a sequence in a closed, bounded set C_2 in $R_{\lambda_1} \times R_{\lambda_2}$, there exists a subsequence $\{(x_{\lambda_1}^{(2,n)}, x_{\lambda_2}^{(2,n)})\}$ of $\{(x_{\lambda_1}^{(1,n)}, x_{\lambda_2}^{(1,n)})\}$ and a number w_2 such that

$$(x_{\lambda_1}^{(2,n)}, x_{\lambda_2}^{(2,n)}) \to (w_1, w_2) \in C_2 \quad \text{as } n \to \infty.$$

In general, since the sequence $\{(x_{\lambda_1}^{(k-1,n)}, \cdots, x_{\lambda_k}^{(k-1,n)})\}$ is a subset of C_k for all $n \geq k$, and since C_k is a closed, bounded subset of $R_{\lambda_1} \times \cdots \times R_{\lambda_k}$, there exists a real number w_k and a subsequence $\{(x_{\lambda_1}^{(k,n)}, \cdots, x_{\lambda_k}^{(k,n)})\}$ which converges to $(w_1, \cdots, w_k)$ in C_k. Now consider the sequence $\{\mathbf{x}^{(n,n)}\}$. The λ_kth coordinates of $\{\mathbf{x}^{(n,n)}\}$ form a subsequence of the λ_kth coordinate of $\mathbf{x}^{(k,n)}$, and consequently converge to w_k. Consider any point in $\Omega = \times_{\lambda \in \Lambda} R_\lambda$ whose λ_kth coordinate is w_k; call this point $\mathbf{w}$. Since $(w_1, \cdots, w_k) \in C_k$ for every k, then $\mathbf{w} \in D_n$ for every n. Hence $\mathbf{w} \in A_n$ for every n, or $\mathbf{w} \in \cap_{n=1}^{\infty} A_n$. Hence $\cap_{n=1}^{\infty} A_n \neq \phi$, giving us a contradiction to the assumption that $A_n \to \phi$ as $n \to \infty$. By the extension theorem quoted earlier there exists one and only one probability P defined over $\mathcal{G}$, the smallest sigma field of subsets containing the field $\mathcal{G}_0$, such that $P(A) = P_0(A)$ for all $A \in \mathcal{G}_0$. Now for any

$$\omega = (\cdots, x_\lambda, \cdots) \in \Omega = \times_{\lambda \in \Lambda} R_\lambda,$$

let us define $X_\lambda(\omega) = x_\lambda$. Thus

$$[X_\lambda \leq x] = \{\omega = (\cdots, x_\lambda, \cdots) \mid x_\lambda \leq x\} \in \mathcal{G}_0 \subset \mathcal{G},$$

so X_λ is a random variable. Further,

$$P(\cap_{j=1}^{n}[X_{\lambda_j} \leq x_j]) = P_0((\times_{j=1}^{n}(-\infty, x_j])$$
$$\times (\times \{R_\lambda \mid \lambda \in \Lambda \setminus \{\lambda_1, \cdots, \lambda_n\}\}))$$
$$= F_{\lambda_1, \cdots, \lambda_n}(x_1, \cdots, x_n),$$

where $(-\infty, x_j] \subset R_{\lambda_j}$. This concludes the proof of the theorem.

It should be noted that $\times_{\lambda \in \Lambda} R_\lambda$ is really the set of all real-valued functions defined over Λ. If we denote this set by $\mathcal{F}_\Lambda$, then $\mathcal{A}$ is the sigma field generated by all sets of the form $\{ f \in \mathcal{F}_\Lambda \mid f(\lambda_i) \leq x_i, 1 \leq i \leq n \}$ for

$$\{\lambda_1, \cdots, \lambda_n\} \subset \Lambda \qquad \text{and} \qquad \{x_1, \cdots, x_n\} \subset (-\infty, +\infty),$$

and $F_{\lambda_1, \cdots, \lambda_n}(x_1, \cdots, x_n)$ is just the probability of this set.

This theorem has many applications and is basic to all the strong limit theorems in probability theory. For example, one might see a statement of the following form: let $\{X_n\}$ be a sequence of random variables such that the joint distribution function of $X_{k_1}, \cdots, X_{k_n}$ is

$$F_{X_{k_1}, \cdots, X_{k_n}}(x_1, \cdots, x_n) = \prod_{j=1}^{n} F(x_{k_j}),$$

where F is some fixed univariate distribution function. The question, of course, is: does there exist a probability space $(\Omega, \mathcal{A}, P)$ with random variables $\{X_n\}$ defined over Ω such that the joint distribution function of every finite subset is as given above? The answer is that since $\prod_{j=1}^{n} F(x_{k_j})$ satisfies conditions (a)–(f) in Theorem 1, then by Theorem 2 we may conclude that there is such a space.

In fact, Theorem 2 is precisely what distinguishes probability theory from other branches of analysis and in particular from measure theory. The statement the theorem allows us to make is this: if $\{X_\lambda, \lambda \in \Lambda\}$ is any set of random variables over some probability space $\{\Omega, \mathcal{A}, P\}$ and if $B \in \sigma\{X_\lambda, \lambda \in \Lambda\}$, then $P(B)$ is determined for all such B by the class of all joint distribution functions of all finite subsets of $\{X_\lambda, \lambda \in \Lambda\}$.

EXERCISES

1. Let $\mathcal{A}_0$ be a field of subsets (that is, closed under finite unions and complementation) of Ω, and let P_0 be a finitely additive probability over $\mathcal{A}_0$, that is,

$$P_0(A) \geq 0$$

for all $A \in \mathcal{A}_0$, $P_0(\Omega) = 1$, and

$$P_0\left(\bigcup_{k=1}^{n} A_k \right) = \sum_{k=1}^{n} P_0(A_k)$$

for every finite collection $A_1, \cdots, A_n$ of disjoint events in $\mathcal{A}_0$. Prove that P_0 is a measure over $\mathcal{A}_0$ if and only if $A_n \in \mathcal{A}_0$, $A_n \supset A_{n+1}$ for all n and $A_n \to \phi$ as $n \to \infty$ imply $P_0(A_n) \to 0$ as $n \to \infty$.

2.4. Expectation

One of the basic ingredients of probability theory is the concept of expectation of a random variable or of its distribution. In this section expectation is defined and its computation is explored.

Definition: If X is a random variable, its expectation EX, or $E(X)$, is defined to be the abstract Lebesgue integral of X with respect to the measure P; that is, $EX = \int X \, dP$, provided this integral exists and is finite. In such a case we shall sometimes write $X \in L_1(\Omega, \alpha, P)$.

Accordingly, if X and Y are in $L_1(\Omega, \alpha, P)$, then $E(aX + bY) = aEX + bEY$ for real a and b, and if $P[X \geq 0] = 1$, then $EX \geq 0$. All properties of the abstract Lebesgue integral over finite measure spaces hold.

Lemma 1. If B is a Borel set in $(-\infty, +\infty)$, and if X is a random variable with distribution function F, then

$$P[X \in B] = \int_B dF,$$

where the integral is the Lebesgue-Stieltjes integral.

Proof: Let μ_F denote the Lebesgue-Stieltjes measure over $(-\infty, +\infty)$ determined by F. For every set B of the form $B = \cup_{i=1}^{n}(a_i, b_i]$ where $b_i \leq a_{i+1}$ and n is finite, we have

$$\mu_F(B) = \sum_{i=1}^{n}(F(b_i) - F(a_i)) = \sum_{i=1}^{n}P[a_i < X \leq b_i]$$

$$= P(\cup_{i=1}^{n}[a_i < X \leq b_i]) = PX^{-1}(B).$$

Now all sets of the form of B just given are a field of subsets of $(-\infty, +\infty)$ which generate the sigma field of Borel sets. By the above equality, PX^{-1} is a measure over the field since μ_F is. Thus by the extension theorem quoted in the proof of Theorem 2 in Section 2.3, the measures μ_F and PX^{-1} are uniquely extended to (and are therefore equal over) the sigma field of Borel sets in $(-\infty, +\infty)$; that is,

$$\int_B d\mu_F\left(= \int_B dF(x)\right) = PX^{-1}(B) = P[X \in B]$$

for all Borel sets B. Q.E.D.

The following theorem is a key theorem for expectation in probability theory.

Theorem 1. If f is a Borel-measurable function defined over $(-\infty, +\infty)$, and if X is a random variable, then the expectation of the random variable $f(X)$ exists if and only if

$$\int |f(x)| \, dF_X(x) < \infty,$$

in which case

$$Ef(X) = \int f(x) \, dF_X(x),$$

where the integrals exhibited are Lebesgue-Stieltjes integrals.

Proof: Assume that the range of f is countable, that is, there is a sequence of real numbers $\{y_n\}$ and a corresponding sequence $\{B_n\}$ of Borel sets in $(-\infty, +\infty)$ such that $f = \sum y_n I_{B_n}$. Then

$$f(X) = \sum y_n I_{[X \in B_n]}, \qquad \text{and} \qquad f(X) \in L_1(\Omega, \mathfrak{a}, P)$$

if and only if

$$\sum |y_n| P[X \in B_n] < \infty.$$

By Lemma 1,

$$P[X \in B_n] = \mu_F(B_n),$$

and hence

$$f(X) \in L_1(\Omega, \mathfrak{a}, P)$$

if and only if

$$\sum |y_n| \mu_F(B_n) < \infty,$$

where μ_F is the Lebesgue-Stieltjes measure determined by F, in which case

$$Ef(X) = \sum y_n P[X \in B_n] = \sum y_n \mu_F(B_n)$$
$$= \int f(x) \, dF_X(x).$$

By the usual method of approximating f by a sequence of countably-valued Borel-measurable functions we obtain the conclusion.

Corollary 1. If, in Theorem 1, f is continuous, then $Ef(X)$ exists if and only if the (improper) Stieltjes integral

$$\int_{-\infty}^{\infty} |f(x)| \, dF_X(x)$$

is finite, in which case the improper Stieltjes integral

$$\int_{-\infty}^{\infty} f(x) \, dF(x)$$

exists and is equal to $Ef(X)$.

Proof: This is a well-known property relating the two integrals.

Corollary 2. If X is a random variable with distribution function F, then EX exists if and only if the two (improper) Stieltjes integrals

$$\int_{0}^{\infty} x \, dF(x) \qquad \text{and} \qquad \int_{-\infty}^{0} x \, dF(x)$$

are finite, in which case

$$EX = \int_{-\infty}^{\infty} x \, dF(x).$$

Proof: This is a particular case of Corollary 1.

Corollary 3. Let X be a random variable with distribution function F. Then EX exists if and only if the two improper Riemann integrals

$$\int_{0}^{\infty} (1 - F(x)) \, dx \qquad \text{and} \qquad \int_{-\infty}^{0} F(x) \, dx$$

are finite, in which case

$$EX = \int_0^\infty (1 - F(x)) \, dx - \int_{-\infty}^0 F(x) \, dx.$$

Proof: This follows from Corollary 2 and by integration by parts.

Corollary 4. If X is discrete, if $P[X = x_n] = p_n$ where $\{x_n\}$ are distinct, if $\sum p_n = 1$, and if EX exists, then $EX = \sum x_n p_n$.

Corollary 5. If X is a random variable with an absolutely continuous distribution function with density f, and if EX exists, then

$$EX = \int_{-\infty}^\infty x f(x) \, dx,$$

where the integral is a Lebesgue integral.

Proof: This follows from Corollary 2 and well-known properties of Radon-Nikodym derivatives.

Examples of computations of expectations for some distributions introduced in Section 2.1 are given below:

(a) Normal distribution, $\mathfrak{N}(\mu, \sigma^2)$. In this case the random variable X has a distribution function F whose density is

$$f(x) = (2\pi\sigma^2)^{-1/2} \exp\left[-(x - \mu)^2/2\sigma^2\right], \qquad -\infty < x < \infty,$$

where $\sigma^2 > 0$ and $-\infty < \mu < \infty$. Thus, by Corollary 5 above,

$$EX = \int_{-\infty}^\infty x f(x) \, dx = (2\pi\sigma^2)^{-1/2} \int_{-\infty}^\infty x \exp\left[\frac{-(x - \mu)^2}{2\sigma^2}\right] dx$$

$$= (2\pi\sigma^2)^{-1/2} \left\{ \int_{-\infty}^\infty (x - \mu) \exp\left[\frac{-(x - \mu)^2}{2\sigma^2}\right] dx \right.$$

$$\left. + \mu \int_{-\infty}^\infty \exp\left[\frac{-(x - \mu)^2}{2\sigma^2}\right] dx \right\}$$

$$= \mu.$$

(b) Binomial distribution, $B(n, p)$. In this case the random variable X has a discrete distribution given by

$$P[X = k] = \binom{n}{k} p^k (1 - p)^{n-k}, \qquad k = 0, 1, \cdots, n,$$

where $0 < p < 1$. Thus, by Corollary 4 above,

$$EX = \sum_{k=0}^{n} k \binom{n}{k} p^k (1 - p)^{n-k} = np.$$

(c) The Poisson distribution. In this case, $P[X = n] = e^{-\lambda} \lambda^n / n!$, $n = 0, 1, 2, \cdots$, where $\lambda > 0$, and $EX = \sum n e^{-\lambda} \lambda^n / n! = \lambda$.

(d) Cantor distribution (see Section 2.1) In this case F is continuous and is symmetric about $\frac{1}{2}$; that is, $P[X > x + \frac{1}{2}] = P[X \le -x + \frac{1}{2}]$. Hence by Corollary 3 above,

$$E(X - \tfrac{1}{2}) = \int_0^\infty P[X - \tfrac{1}{2} > x] \, dx - \int_{-\infty}^0 P[X - \tfrac{1}{2} \le x] \, dx$$

$$= \int_0^{1/2} P[X > x + \tfrac{1}{2}] \, dx - \int_{-1/2}^0 P[X \le x + \tfrac{1}{2}] \, dx$$

$$= 0,$$

and hence $EX = E(X - \tfrac{1}{2}) + E(\tfrac{1}{2}) = \tfrac{1}{2}$.

(e) Cauchy distribution. In this case the distribution function F has a density $f(x) = 1/\pi(1 + x^2)$, $-\infty < x < \infty$. However,

$$\int_{-\infty}^\infty x f(x) \, dx$$

does not exist in this case, so a random variable with a Cauchy distribution does not have an expectation.

The following definition will be needed in the next section.

Definition: If X and Y are random variables, we define $X + iY$ by

$$(X + iY)(\omega) = X(\omega) + iY(\omega);$$

we shall call this a complex-valued random variable. Its expectation is said to exist if and only if EX and EY exist, in which case we define $E(X + iY) = E(X) + iE(Y)$.

The moments of random variables play an important role in probability theory; these moments can now be defined with the help of Theorem 1.

Definition: The nth moment of a random variable X with distribution function F is defined to be

$$E(X^n) = \int_{-\infty}^{\infty} x^n \, dF(x),$$

provided this expectation exists. The nth central moment of X is defined to be

$$E(X - EX)^n = \int_{-\infty}^{\infty} (x - EX)^n \, dF(x),$$

provided this expectation exists. In both cases we might write

$$X \in L_n(\Omega, \mathcal{Q}, P).$$

The *variance* of X is the second central moment of X and is denoted by Var X.

Very easy computations yield the following facts: (a) if the distribution of X is $\mathfrak{N}(\mu, \sigma^2)$, then Var $X = \sigma^2$, (b) if the distribution of X is $B(n, p)$, then Var X is $np(1 - p)$, and (c) if the distribution of X is $\mathcal{P}(\lambda)$, then Var $X = \lambda$.

The following theorem and, in particular, the corollary are among the most frequently used tools in probability theory.

Theorem 2. If X is a random variable and if $E \mid X \mid^r < \infty$ for $r > 0$ not necessarily an integer, then $P[\mid X \mid \geq \epsilon] \leq E \mid X \mid^r / \epsilon^r$ for every $\epsilon > 0$.

Proof: We observe, letting $F(x) = P[X \leq x]$, that

$$E \mid X \mid^r = \int_{-\infty}^{\infty} \mid x \mid^r dF(x) = \int_{\mid x \mid < \epsilon} \mid x \mid^r dF(x) + \int_{\mid x \mid \geq \epsilon} \mid x \mid^r dF(x)$$

$$\geq \int_{\mid x \mid \geq \epsilon} \mid x \mid^r dF(x) \geq \epsilon^r \int_{\mid x \mid \geq \epsilon} dF(x)$$

$$= \epsilon^r P[\mid X \mid \geq \epsilon],$$

from which we obtain the inequality.

Corollary to Theorem 2 (Chebishev's Inequality). If $X \in L_2(\Omega, \alpha, P)$, then

$$P[\mid X - EX \mid \geq \epsilon] \leq \operatorname{Var} X / \epsilon^2 \qquad \text{for every } \epsilon > 0.$$

Proof: This is a special case of Theorem 2.

EXERCISES

1. Prove: If $\{B_n\}$ is a sequence of disjoint Borel sets in $(-\infty, +\infty)$, if $\{y_n\}$ is a sequence of real numbers, if $f = \sum y_n I_{B_n}$, and if X is a random variable, then

$$f(X) = \sum y_n I_{[X \in B_n]}.$$

2. Prove: if X is a random variable with distribution function F, and if $X \in L_1(\Omega, \alpha, P)$, then

$$E \mid X \mid = \int \mid x \mid dF(x) = \int_{-\infty}^{0} F(x)\, dx + \int_{0}^{\infty} (1 - F(x))\, dx.$$

3. Let f be a Borel-measurable function defined over $E^{(n)}$, and let $X_1, \cdots, X_n$ be n random variables with joint distribution function F. Then $Ef(X_1, \cdots, X_n)$ exists if and only if

$$\int \cdots \int \mid f(x_1, \cdots, x_n) \mid dF(x_1, \cdots, x_n) < \infty,$$

in which case

$$Ef(X_1, \cdots, X_n) = \int \cdots \int f(x_1, \cdots, x_n) \, dF(x_1, \cdots, x_n),$$

where both integrals are Lebesgue-Stieltjes integrals.

4. In Problem 3 show that in case F is absolutely continuous with joint density φ, then

$$Ef(X_1, \cdots, X_n) = \int \cdots \int f(x_1, \cdots, x_n) \varphi(x_1, \cdots, x_n) \, dx_1 \cdots dx_n,$$

where the integral is a Lebesgue integral.

5. If $\mathbf{X} = (X_1, \cdots, X_n)$ is a discrete n-dimensional random variable, and if h is a Borel-measurable function defined over $E^{(n)}$, then

$$Eh(X_1, \cdots, X_n) = \sum_{\mathbf{x} \in E^{(n)}} h(\mathbf{x}) P[\mathbf{X} = \mathbf{x}],$$

provided this series is absolutely convergent.

6. Let X be a random variable whose distribution function F is continuous. Evaluate $EF(X)$.

7. Evaluate Var X when the distribution of X is (a) $B(n, p)$, (b) $\mathfrak{N}(\mu, \sigma^2)$, and (c) $\mathcal{P}(\lambda)$.

8. Prove: if $X \in L_2(\Omega, \mathcal{C}, P)$, then

$$\mathrm{Var}\ (X) = E(X^2) - (EX)^2.$$

9. If X and Y are in $L_2(\Omega, \mathcal{C}, P)$, then the correlation coefficient of X and Y, $\rho(X, Y)$, is defined by

$$\rho(X, Y) = \frac{E[(X - EX)(Y - EY)]}{\sqrt{\mathrm{Var}\ (X)\ \mathrm{Var}\ (Y)}}.$$

Prove that $-1 \le \rho(X, Y) \le 1$, and find conditions under which equality in each of the two inequalities is achieved.

2.5. Characteristic Functions

Probability theory is distinguished from measure theory in that its theorems are concerned with distribution functions or properties that are determined by distribution functions. Distribution functions are not always easy to work with directly, but they can frequently be handled by means of their characteristic functions. This section is devoted to some basic properties of characteristic functions.

Definition: If X is a random variable, its characteristic function, $f(u)$, is defined by

$$f(u) = E \exp \{iuX\}, \quad -\infty < u < \infty.$$

If F is a distribution function, its characteristic function is defined to be the Fourier-Stieltjes transform

$$f(u) = \int_{-\infty}^{\infty} \exp (iux) \, dF(x), \quad -\infty < u < \infty.$$

Thus, a characteristic function is a complex-valued function of a real variable.

Proposition 1. If X is a random variable whose distribution function is F, then the characteristic functions of X and F are well-defined and are equal.

Proof: Since $\cos ux$ and $\sin ux$ are bounded continuous functions of x (for each u), they are Borel measurable and bounded so

$$E[\exp (iuX)] = E \cos uX + iE \sin uX$$

exists. Equality of the two characteristic functions follows from Theorem 1 in Section 2.4.

Definition: If $\mathbf{X} = (X_1, \cdots, X_n)$ is an n-dimensional random variable, its joint characteristic function is defined to be

$$f(\mathbf{u}) = E \exp \{i(\mathbf{u}, \mathbf{X})\}$$

for all

$$\mathbf{u} = (u_1, \cdots, u_n) \in E^{(n)},$$

where

$$(\mathbf{u}, \mathbf{X}) = \sum_{j=1}^{n} u_j X_j .$$

If $F(x_1, \cdots, x_n)$ is a multivariate distribution function, its characteristic function is defined by

$$f(\mathbf{u}) = \int \cdots \int \exp [i(\mathbf{u}, \mathbf{x})] \, dF(\mathbf{x}) \quad \text{for all} \quad \mathbf{u} \in E^{(n)}.$$

As in Proposition 1, if $F(x_1, \cdots, x_n)$ is the distribution function of $\mathbf{X} = (X_1, \cdots, X_n)$, then

$$E \exp \{i \sum_{j=1}^{n} u_j X_j\} = \int_{-\infty}^{\infty} \cdots \int_{-\infty}^{\infty} \exp \{i \sum_{j=1}^{n} u_j x_j\} \, dF(x_1, \cdots, x_n).$$

Proposition 2. If $f(\mathbf{u})$ is the characteristic function of a distribution function $F(\mathbf{x})$, $\mathbf{x} \in E^{(n)}$, then

(a) $f(\mathbf{0}) = 1$, where $\mathbf{0} = (0, 0, \cdots, 0)$,
(b) $|f(\mathbf{u})| \leq 1$ for all $\mathbf{u} \in E^{(n)}$, and
(c) f is uniformly continuous over $E^{(n)}$.

Proof: (a) $f(\mathbf{0}) = \int \cdots \int e^0 \, dF(\mathbf{x}) = 1.$

(b) $|f(\mathbf{u})| = |\int \cdots \int \exp [i(\mathbf{u}, \mathbf{x})] \, dF(\mathbf{x})|$

$$\leq \int \cdots \int |\exp [i(\mathbf{u}, \mathbf{x})]| \, dF(\mathbf{x})$$

$$= \int \cdots \int dF(\mathbf{x}) = 1.$$

(c) For every $\mathbf{h} \in E^{(n)}$ and $\mathbf{u} \in E^{(n)}$, we observe that

$$|f(\mathbf{u} + \mathbf{h}) - f(\mathbf{u})| = |\int \{\exp [i(\mathbf{u} + \mathbf{h}, \mathbf{x})]$$

$$- \exp [i(\mathbf{u}, \mathbf{x})]\} \, dF(\mathbf{x})| \leq \int |\exp [i(\mathbf{h}, \mathbf{x})] - 1| \, dF(\mathbf{x}),$$

which does not depend on $\mathbf{u}$. It remains to show that this last term tends to zero as $\mathbf{h} \to \mathbf{0}$. Since $2 \geq |\exp [i(\mathbf{h}, \mathbf{x})] - 1| \to 0$ as $\mathbf{h} \to \mathbf{0}$ for every $\mathbf{x} \in E^{(n)}$, the Lebesgue dominated convergence theorem applies, which yields the conclusion.

Proposition 3. If $\mathbf{X}$ is an n-dimensional random variable with characteristic function $f(\mathbf{u})$, and if $\mathbf{Y}$ is an n-dimensional random variable defined by

$$Y_i = a_i + b_i X_i, \qquad 1 \le i \le n,$$

where a_i and b_i are real numbers, then the characteristic function of $\mathbf{Y}$, $g(\mathbf{u})$, is given by

$$g(\mathbf{u}) = \exp\left[i(\mathbf{a}, \mathbf{u})\right] f(b_1 u_1, \cdots, b_n u_n),$$

where

$$\mathbf{a} = (a_1, \cdots, a_n).$$

Proof: By a direct computation we obtain

$$g(\mathbf{u}) = E \exp\left\{ i \sum_{j=1}^{n} u_j Y_j \right\}$$

$$= E \exp\left\{ i \sum_{j=1}^{n} u_j a_j + i \sum_{j=1}^{n} u_j b_j X_j \right\}$$

$$= \exp\left\{ i \sum_{j=1}^{n} u_j a_j \right\} f(u_1 b_1, \cdots, u_n b_n),$$

which concludes the proof.

Proposition 4. If $f(u)$ is the characteristic function of $F(x)$, then $\overline{f(u)}$ is the characteristic function of $G(x)$, where $G(x) = 1 - F(-x - 0)$.

Proof: An easy direct computation yields

$$\overline{f(u)} = \overline{\int_{-\infty}^{\infty} \exp(iux)\, dF(x)} = \int_{-\infty}^{\infty} \exp(-iux)\, dF(x)$$

$$= \int_{-\infty}^{\infty} \exp(iux)\, d(-F(-x)) = \int_{-\infty}^{\infty} \exp(iux)\, dG(x).$$

We next compute the characteristic functions of frequently encountered distributions.

(**a**) *Binomial distribution*, $B(n, p)$. Here

$$P[X = k] = \binom{n}{k} p^k (1 - p)^{n-k}, \qquad 0 \le k \le n,$$

and so

$$E[\exp (iuX)] = \sum_{k=0}^{n} \exp (iuk) \binom{n}{k} p^k (1 - p)^{n-k},$$

$$= ((1 - p) + pe^{iu})^n.$$

(**b**) *Poisson distribution*, $\mathcal{P}(\lambda)$. Here we have

$$P[X = n] = \exp (-\lambda)\lambda^n/n!, \qquad n = 0, 1, 2, \cdots.$$

Thus,

$$E[\exp (iuX)] = \sum_{n=0}^{\infty} \exp (iun)P[X = n]$$

$$= \sum_{n=0}^{\infty} \exp (iun) \exp (-\lambda)\lambda^n/n! = \exp [\lambda(e^{iu} - 1)].$$

(**c**) *Normal distribution*, $\mathfrak{N}(\mu, \sigma^2)$. We first compute the characteristic function in the special case when the distribution of X is

$$P[X \le x] = (2\pi)^{-1/2} \int_{-\infty}^{x} \exp (-t^2/2) \, dt,$$

that is, $\mathfrak{N}(0, 1))$, so X has a density

$$\varphi(x) = (2\pi)^{-1/2} \exp (-x^2/2).$$

The characteristic function $f(u)$ is given by

$$f(u) = (2\pi)^{-1/2} \int_{-\infty}^{\infty} \exp (iux - x^2/2) \, dx$$

$$= \exp (-u^2/2)(2\pi)^{-1/2} \int_{-\infty}^{\infty} \exp [-(x - iu)^2/2] \, dx.$$

In order to evaluate this integral, consider the contour integral

$$\int_R \exp{(-z^2/2)}\, dz$$

around the rectangle whose vertices are $\pm K + i0$, $\pm K - iu$. Since $\exp{(-z^2/2)}$ is entire, we have

$$\int_R \exp{(-z^2/2)}\, dz = \int_{-K}^{K} \exp{(-x^2/2)}\, dx$$

$$+ \int_{+K}^{-K} \exp{[-(x - iu)^2/2]}\, dx$$

$$+ \int_0^{-u} \exp{[-(K + it)^2/2]}\, dt$$

$$+ \int_{-u}^{0} \exp{[-(-K + it)^2/2]}\, dt = 0.$$

Each of the last two integrals is bounded in modulus by $|u| \exp{(-K^2/2)}$, which tends to zero as $K \to \infty$. Letting $K \to \infty$, and remembering that

$$(2\pi)^{-1/2} \int_{-\infty}^{\infty} \exp{(-x^2/2)}\, dx = 1,$$

we obtain

$$f(u) = \exp{(-u^2/2)}.$$

In the general case when the distribution of X is $\mathfrak{N}(\mu,\ \sigma^2)$, let $Z = (X - \mu)/\sigma$. Then

$$P[Z \le z] = P\left[\frac{X - \mu}{\sigma} \le z\right] = P[X \le \sigma z + \mu]$$

$$= (2\pi\sigma^2)^{-1/2} \int_{-\infty}^{\sigma z + \mu} \exp{\{-(t - \mu)^2/2\sigma^2\}}\, dt$$

$$= (2\pi)^{-1/2} \int_{-\infty}^{z} \exp{\{-\tau^2/2\}}\, d\tau;$$

that is, the distribution of Z is $\mathfrak{N}(0, 1)$. By the computation just completed, the characteristic function of Z is

$$f_Z(u) = \exp\{-u^2/2\}.$$

Since $X = \sigma Z + \mu$, we obtain by Proposition 3 that

$$f_X(u) = \exp(i\mu u)f_Z(\sigma u)$$

$$= \exp\{i\mu u - \sigma^2 u^2/2\}.$$

This characteristic function plays an important role from here on.

(d) *Cauchy distribution.* In this case X has a density

$$\varphi(x) = 1/\pi(1 + x^2), \qquad -\infty < x < \infty,$$

and its characteristic function is

$$f(u) = \int_{-\infty}^{\infty} \frac{\exp(iux)}{\pi(1 + x^2)}\, dx.$$

We evaluate this integral in two cases. Case: $u > 0$. Consider the contour $C: -R \le x \le R$, $z = Re^{i\theta}$, $0 \le \theta \le \pi$. For $R > 1$, consider

$$\int_C \frac{\exp(iuz)}{\pi(1 + z^2)}\, dz$$

$$= \int_{-R}^{R} \frac{\exp(iux)}{\pi(1 + x^2)}\, dx + \int_0^{\pi} \frac{\exp\{iuR\cos\theta - uR\sin\theta\}}{\pi[1 + R^2 \exp(i2\theta)]}\, iRe^{i\theta}\, d\theta$$

$$= 2\pi i \frac{\exp(iuz)}{\pi(z + i)}\bigg|_{z=i} = e^{-u}.$$

As $R \to +\infty$, the first of the two integrals converges to $f(u)$. Since $u > 0$, the modulus of the integrand in the second integral is bounded by $R/\pi(R^2 - 1)$, and thus the second integral tends to 0 as $R \to \infty$. In Case $u < 0$, we consider the contour $C': +R \ge x \ge -R$, $z = Re^{i\theta}$, $-\pi \le \theta \le 0$. In this case we obtain in a similar manner that $f(u) = e^u$. Since $f(0) = 1$, then whatever u may be, we obtain

$$f(u) = \exp(-|u|), \qquad -\infty < u < \infty.$$

(e) *Uniform distribution* over $[-1, 1]$. Here the random variable X has a density

$$\varphi(x) = \begin{cases} \frac{1}{2} & \text{if} \quad -1 \leq x \leq 1 \\ 0 & \text{if} \quad |x| > 1. \end{cases}$$

Thus,

$$f(u) = \int_{-\infty}^{\infty} \exp(iux)\varphi(x) \, dx = \frac{1}{2} \int_{-1}^{1} \exp(iux) \, dx = \frac{\sin u}{u}.$$

The principal result that we wish to obtain in this section is that there is a one-to-one correspondence between the set $\mathfrak{D}$ of all distribution functions and the set $\mathfrak{C}$ of all characteristic functions. In order to do this we first need two lemmas.

Lemma 1. If $\varphi \in L_1(-\infty, +\infty)$, if φ is continuous everywhere, and if

$$\hat{\varphi}(u) = \int_{-\infty}^{\infty} \varphi(x) \exp(iux) \, dx \in L_1(-\infty, +\infty),$$

then

$$\varphi(x) = \frac{1}{2\pi} \int_{-\infty}^{\infty} \hat{\varphi}(u) \exp(-iux) \, du \quad \text{for all} \quad x.$$

Proof: For $\sigma^2 \geq 0$ define Φ by

$$\Phi(\sigma^2) = \frac{1}{2\pi} \int_{-\infty}^{\infty} \hat{\varphi}(u) \exp(-iux) \exp(-\sigma^2 u^2/2) \, du.$$

By hypothesis, $\hat{\varphi} \in L_1(-\infty, +\infty)$, so the integral above is absolutely convergent, and the integrand is bounded in modulus by $|\hat{\varphi}(u)|$ for all $\sigma^2 \geq 0$. Thus, by the Lebesgue dominated convergence theorem,

$$\lim_{\sigma^2 \to 0} \Phi(\sigma^2) = \frac{1}{2\pi} \int_{-\infty}^{\infty} \hat{\varphi}(u) \exp(-iux) \, du.$$

By Fubini's theorem we may write

$$\Phi(\sigma^2) = \frac{1}{2\pi} \int_{-\infty}^{\infty} \varphi(t) \left(\int_{-\infty}^{\infty} \exp \left[iu(t - x) - \frac{\sigma^2 u^2}{2} \right] du \right) dt.$$

The inside integral is clearly equal to $\sqrt{2\pi/\sigma^2}$ times the characteristic function evaluated at $t - x$ of the normal distribution $\mathfrak{N}(0, 1/\sigma^2)$, and so by the previous computation of this characteristic function we have

$$\Phi(\sigma^2) = (2\pi\sigma^2)^{-1/2} \int_{-\infty}^{\infty} \varphi(t) \exp \left[\frac{-(t - x)^2}{2\sigma^2} \right] dt.$$

Since φ is continuous at x, then for arbitrary $\epsilon > 0$ there exists a $\delta > 0$ such that if $| t - x | < \delta$, then $| \varphi(t) - \varphi(x) | < \epsilon/3$. Having selected $\delta > 0$ (fixed), let $\sigma^2 > 0$ be selected so small that

$$\frac{\exp(-\delta^2/2\sigma^2)}{\sqrt{2\pi\sigma^2}} < \epsilon/3 \max \left\{ 1, \int | \varphi(\tau) | d\tau \right\}$$

and

$$\frac{\sigma^2 | \varphi(x) |}{\delta^2} < \epsilon/3.$$

Now

$$\Phi(\sigma^2) = (2\pi\sigma^2)^{-1/2} \int_{-\infty}^{\infty} \{ (\varphi(t) - \varphi(x)) + \varphi(x) \}$$

$$\times \exp \left[-(t - x)^2/2\sigma^2 \right] dt$$

$$= (2\pi\sigma^2)^{-1/2} \int_{-\infty}^{\infty} (\varphi(t) - \varphi(x)) \exp \left[-(t - x)^2/2\sigma^2 \right] dt + \varphi(x).$$

Thus,

$$| \Phi(\sigma^2) - \varphi(x) | \leq I_1 + I_2 + I_3,$$

where

$$I_1 = \int_{x-\delta}^{x+\delta} (2\pi\sigma^2)^{-1/2} | \varphi(t) - \varphi(x) | \exp \left[-(t - x)^2/2\sigma^2 \right] dt,$$

$$I_2 = (2\pi\sigma^2)^{-1/2} \int_{-\infty}^{x-\delta} + \int_{x+\delta}^{\infty} | \varphi(t) | \exp \left[-(t - x)^2/2\sigma^2 \right] dt,$$

and

$$I_3 = |\varphi(x)| (2\pi\sigma^2)^{-1/2} \int_{-\infty}^{x-\delta} + \int_{x+\delta}^{\infty} \exp\left[-(t-x)^2/2\sigma^2\right] dt.$$

Now

$$I_1 < (\epsilon/3) \int_{-\infty}^{\infty} (2\pi\sigma^2)^{-1/2} \exp\left[-(t-x)^2/2\sigma^2\right] dt = \epsilon/3.$$

Also

$$I_2 \leq \exp\left(-\delta^2/2\sigma^2\right) (2\pi\sigma^2)^{-1/2} \int_{-\infty}^{\infty} |\varphi(t)| \, dt < \epsilon/3.$$

In order to find an upper bound for I_3, we note that if X is a random variable whose distribution function is $\mathfrak{N}(x, \sigma^2)$, then

$$I_3 = |\varphi(x)| P[|X - x| \geq \delta].$$

By Chebishev's inequality,

$$I_3 \leq |\varphi(x)| \sigma^2/\delta^2 < \tfrac{1}{3}\epsilon.$$

Thus, for $\sigma^2 > 0$ sufficiently small, we have $|\Phi(\sigma^2) - \varphi(x)| < \epsilon$. Thus, $\lim_{\sigma^2 \downarrow 0} \Phi(\sigma^2) = \varphi(x)$, which proves the lemma.

Lemma 2. Let $-\infty < a < b < \infty$, $\epsilon > 0$, define

$$g^{(\epsilon)}(x) = \begin{cases} 1 & \text{if } a \leq x \leq b \\ 0 & \text{if } x \geq b + \epsilon \quad \text{or} \quad x \leq a - \epsilon \\ (b - x + \epsilon)/\epsilon & \text{if } b < x < b + \epsilon \\ (x - a + \epsilon)/\epsilon & \text{if } a - \epsilon < x < a, \end{cases}$$

and denote

$$\hat{g}^{(\epsilon)}(u) = \int_{-\infty}^{\infty} g^{(\epsilon)}(x) \exp(iux) \, dx.$$

Then $|\hat{g}^{(\epsilon)}(u)| = 0(1/u^2)$ as $|u| \to \infty$.

Proof: Let $c = (a + b)/2$ and $\kappa = (b - a)/2$. Let $y = x - c$ or $x = y + c$. Then

$$\hat{g}^{(\epsilon)}(u) = e^{iuc} \int_{-\kappa-\epsilon}^{\kappa+\epsilon} g^{(\epsilon)}(y + c) e^{iuy} \, dy$$

$$= 2e^{iuc} \left\{ \int_0^\kappa \cos ux \, dx + \int_\kappa^{\kappa+\epsilon} \frac{\kappa + \epsilon - x}{\epsilon} \cos ux \, dx \right\}$$

$$= 2e^{iuc} \left\{ \frac{\sin u\kappa}{u} + \int_0^\epsilon \frac{\epsilon - z}{\epsilon} \cos u(z + \kappa) \, dz \right\}$$

$$= \frac{2e^{iuc}}{u\epsilon} \int_0^\epsilon \sin u(z + \kappa) \, dz$$

$$= (2e^{iuc}/u^2\epsilon)(-\cos u(\epsilon + \kappa) + \cos u\kappa)$$

$$= 0(1/u^2) \quad \text{as} \quad |u| \to \infty. \quad \text{Q.E.D.}$$

We are now in a position to prove the uniqueness theorem for multivariate distribution functions.

Theorem 1. Let F and G be two joint distribution functions defined over $E^{(n)}$ with corresponding characteristic functions $f(\mathbf{u})$ and $g(\mathbf{u})$. If $f = g$, then $F = G$.

Proof: Let $g_j^{(\epsilon)}(x)$ be defined as $g^{(\epsilon)}(x)$ was in Lemma 2 above, except that a and b are replaced by a_j and b_j, where $-\infty < a_j < b_j < \infty$. Let us denote

$$\hat{g}_j^{(\epsilon)}(u_j) = \int_{-\infty}^\infty g_j^{(\epsilon)}(x_j) \exp(iu_jx) \, dx.$$

By Lemma 2,

$$|\hat{g}_j^{(\epsilon)}(u_j)| = 0(1/u_j^2) \quad \text{as} \quad |u_j| \to \infty,$$

and thus

$$\hat{g}_j^{(\epsilon)}(u_j) \in L_1(-\infty, +\infty).$$

Now $g_j^{(\epsilon)}(x_j)$ is continuous and in $L_1(-\infty, +\infty)$, so by Lemma 1,

$$(1) \qquad g_j^{(\epsilon)}(x_j) = \frac{1}{2\pi} \int_{-\infty}^\infty \hat{g}_j^{(\epsilon)}(u_j) \exp(-iu_jx_j) \, du_j.$$

Let us denote

$$I = \underset{j=1}{\overset{n}{\times}} [a_j, b_j] \subset E^{(n)},$$

$$g_I^{(\epsilon)}(x_1, \cdots, x_n) = \prod_{j=1}^{n} g_j^{(\epsilon)}(x_j),$$

and

$$\hat{g}_I^{(\epsilon)}(u_1, \cdots, u_n) = \int \cdots \int g_I^{(\epsilon)}(x_1, \cdots, x_n)$$

$$\times \exp\left(i \sum_{j=1}^{n} u_j x_j\right) dx_1 \cdots dx_n.$$

By Fubini's theorem and the fact that $g_j^{(\epsilon)} \in L_1(-\infty, +\infty)$ for every j we obtain

$$(2) \qquad \hat{g}_I^{(\epsilon)}(u_1, \cdots, u_n) = \prod_{j=1}^{n} \hat{g}_j^{(\epsilon)}(u_j).$$

Now by (1) and (2) we obtain

$$g_I^{(\epsilon)}(\mathbf{x}) = \prod_{j=1}^{n} g_j^{(\epsilon)}(x_j)$$

$$= \prod_{j=1}^{n} \left\{ \frac{1}{2\pi} \int_{-\infty}^{\infty} \hat{g}_j^{(\epsilon)}(u_j) \exp[-iu_j x_j] \, du_j \right\}$$

$$= (2\pi)^{-n} \int_{-\infty}^{\infty} \cdots \int_{-\infty}^{\infty} \left\{ \prod_{j=1}^{n} \hat{g}_j^{(\epsilon)} u_j) \right\}$$

$$\times \exp\left[-i \sum_{j=1}^{n} u_j x_j\right] du_1 \cdots du_n$$

$$= (2\pi)^{-n} \int_{-\infty}^{\infty} \cdots \int_{-\infty}^{\infty} \hat{g}_I^{(\epsilon)}(u_1, \cdots, u_n)$$

$$\times \exp\left[-i \sum_{j=1}^{n} u_j x_j\right] du_1 \cdots du_n.$$

The above integral is absolutely convergent, so by Fubini's theorem and the fact that the characteristic functions of F and G are equal, we get

$$\int g_I^{(\epsilon)}(\mathbf{x} - \mathbf{y})\, dF(\mathbf{y})$$

$$= \int (2\pi)^{-n} \left\{ \int \hat{g}_I^{(\epsilon)}(\mathbf{u}) \exp\left[-i(\mathbf{u}, \mathbf{x} - \mathbf{y})\right] d\mathbf{u} \right\} dF(\mathbf{y})$$

$$= (2\pi)^{-n} \int \left(\int \exp\left[i(\mathbf{u}, \mathbf{y})\right] dF(\mathbf{y}) \right) \exp\left[-i(\mathbf{u}, \mathbf{x})\right]\hat{g}_I^{(\epsilon)}(\mathbf{u})\, d\mathbf{u}$$

$$= (2\pi)^{-n} \int \left(\int \exp\left[i(\mathbf{u}, \mathbf{y})\right] dG(\mathbf{y}) \right) \exp\left[-i(\mathbf{u}, \mathbf{x})\right]\hat{g}_I^{(\epsilon)}(\mathbf{u})\, d\mathbf{u}$$

$$= \int g_I^{(\epsilon)}(\mathbf{x} - \mathbf{y})\, dG(\mathbf{y}).$$

By the Lebesgue dominated convergence theorem we may take limits as $\epsilon \to 0$ and obtain

$$\int \chi_I(\mathbf{x} - \mathbf{y})\, dF(\mathbf{y}) = \int \chi_I(\mathbf{x} - \mathbf{y})\, dG(\mathbf{y}),$$

where $\chi_I(\mathbf{t}) = 1$ if $\mathbf{t} \in I$ and $= 0$ if $\mathbf{t} \notin I$. If μ_F and μ_G denote the Lebesgue-Stieltjes measures determined by F and G, respectively, over $E^{(n)}$, then the above inequality becomes

$$\mu_F[\mathbf{y} \mid \mathbf{x} - \mathbf{y} \in I\} = \mu_G\{\mathbf{y} \mid \mathbf{x} - \mathbf{y} \in I\}$$

for every $\mathbf{x}$. This implies that the μ_F and μ_G measures coincide; that is, $F = G$. Q.E.D.

Definition: A distribution function F is said to be symmetric if $F(x) = 1 - F(-x - 0)$ for all x.

Proposition 5. A distribution function is symmetric if and only if its characteristic function is real.

Proof: Let f be the characteristic function of the distribution function F, and let $G(x) = 1 - F(-x - 0)$. If F is symmetric, that is, if $F = G$, then

$$f(u) = \int_{-\infty}^{\infty} \exp{(iux)} \, dF(x) = \int_{-\infty}^{\infty} \exp{(-iux)} \, dG(x)$$

$$= \int_{-\infty}^{\infty} \exp{(-iux)} \, dF(x) = \overline{f(u)},$$

which implies that f is real. Conversely, if f is real, and since

$$\overline{f(u)} = \int_{-\infty}^{\infty} \exp{(-iux)} \, dF(x) = \int_{-\infty}^{\infty} \exp{(iux)} \, dG(x)$$

and since

$$\overline{f(u)} = f(u),$$

then by the uniqueness theorem (Theorem 1) we obtain that $F = G$, that is, F is symmetric.

So far we have shown that the set of characteristic functions is in one-to-one correspondence with the set of all distribution functions. We next prove an isomorphism theorem.

Theorem 2. If F and G are distribution functions with characteristic functions f and g respectively, then h defined by $h(u) = f(u)g(u)$ is the characteristic function of the distribution function $F * G$.

Proof: By repeated use of the Lebesgue dominated convergence theorem and the definitions of convolution and Stieltjes integral we have

$$\int_{-\infty}^{\infty} \exp\,(iux)\,dF \ast G(x) = \lim_{N \to \infty} \int_{-N}^{N} \exp\,(iux)\,dF \ast G(x)$$

$$= \lim_{N \to \infty} \lim_{n \to \infty} \sum_{k=-N2^n+1}^{N2^n} \exp\left(\frac{iuk}{2^n}\right) \int_{-\infty}^{\infty} \left(F\left(\frac{k}{2^n} - y\right)\right.$$

$$\left. - F\left(\frac{k-1}{2^n} - y\right)\right) dG(y)$$

$$= \lim_{N \to \infty} \lim_{n \to \infty} \sum_{k=-N2^n+1}^{N2^n} \int_{-\infty}^{\infty} \left\{ \exp\left[iu\left(\frac{k}{2^n} - y\right)\right]\left(F\left(\frac{k}{2^n} - y\right)\right.\right.$$

$$\left.\left. - F\left(\frac{k-1}{2^n} - y\right)\right)\right\} \exp\,(iuy)\,dG(y)$$

$$= \int_{-\infty}^{\infty} \left\{ \lim_{N \to \infty} \int_{-N-y}^{N-y} \exp\,(iux)\,dF(x)\right\} \exp\,(iuy)\,dG(y)$$

$$= \int_{-\infty}^{\infty} f(u) \exp\,(iuy)\,dG(y) = f(u)g(u). \qquad \text{Q.E.D.}$$

An easy corollary of this theorem is the fact that convolution is a commutative binary operation.

Corollary: If F and G are distribution functions, then $F \ast G = G \ast F$.

Proof: This is an immediate consequence of Theorems 1 and 2 and the fact that $f(u)g(u) = g(u)f(u)$, where f and g are the characteristic functions of F and G respectively.

EXERCISES

1. Prove: if F, G, and H are distribution functions, then

$$(F \ast G) \ast H = F \ast (G \ast H).$$

2. Prove: if F is the distribution function of the random variable X, then $1 - F(-x - 0)$ is the distribution function of $-X$.

3. Prove that if the distribution function F is $\mathfrak{N}(\mu, \sigma^2)$, then G defined by $G(x) = F(\sigma x + \mu)$ is $\mathfrak{N}(0, 1)$.

4. $\displaystyle\int_{-\infty}^{\infty} \exp\left[i\theta x - 4x^2/\pi\right] dx =$ what constant multiplied by the characteristic function of what distribution function evaluated at what value of the argument?

5. Let F be a distribution function, and define

$$G(x) = 1 - F(-x - 0).$$

Prove that $F * G$ is symmetric.

6. Prove: if $f(u)$ is a characteristic function, so is $|f(u)|^2$.

7. Prove: if f is a characteristic function, then it is positive definite, that is, for every positive integer n, for every n real numbers $u_1, \cdots, u_n$, and for every n complex numbers $z_1, \cdots, z_n$,

$$\sum_{i=1}^{n} \sum_{j=1}^{n} f(u_i - u_j) z_i \bar{z}_j \geq 0.$$

8. Find the characteristic function of the distribution function whose density is

$$f(x) = \tfrac{1}{2} \exp\left(-|x|\right), \qquad -\infty < x < \infty.$$

9. Let F, G be two distribution functions defined over $E^{(n)}$ with corresponding Lebesgue-Stieltjes measures μ_F, μ_G. Prove that if

$$\mu_F\{\mathbf{y} \mid \mathbf{x} - \mathbf{y} \in I\} = \mu_G\{\mathbf{y} \mid \mathbf{x} - \mathbf{y} \in I\}$$

for every $\mathbf{x} \in E^{(n)}$ and for every cell $I \subset E^{(n)}$, then $F = G$.

CHAPTER 3

Stochastic Independence

=====

3.1. Independent Events

One of the most frequently used properties in probability theory is that of independence. This chapter is devoted to an introduction to this concept and some of its properties. We begin with independent events.

Definition: Two events, A and B, are said to be independent if

$$P(AB) = P(A)P(B).$$

Proposition 1. If A and B are independent events, then A and B^c are independent.

Proof: By means of the definition we obtain

$$P(AB^c) = P(A) - P(AB) = P(A) - P(A)P(B)$$
$$= P(A)(1 - P(B)) = P(A)P(B^c),$$

which shows that A and B^c satisfy the definition of independence.

Proposition 2. If A and B are independent events, and if $P(B) > 0$, then $P(A \mid B) = P(A)$.

Proof: By the definition of conditional probability and independence we obtain

$$P(A \mid B) = P(AB)/P(B) = P(A)P(B)/P(B) = P(A).$$

We next consider independence in a class of events.

Definition: Let $\mathcal{B} = \{B_\lambda, \lambda \in \Lambda\}$ be a set of events. These events are said to be independent if for every positive integer n and every n distinct elements $\lambda_1, \cdots, \lambda_n$ in the indexing set Λ, we have

$$P(B_{\lambda_1} \cap \cdots \cap B_{\lambda_n}) = \prod_{j=1}^{n} P(B_{\lambda_j}).$$

Proposition 3. If $\mathcal{C}$ is a collection of independent events, if $A_1, \cdots, A_n, B_1, \cdots, B_m$ are $m + n$ distinct events in $\mathcal{C}$, and if

$$P(B_1 \cap \cdots \cap B_m) > 0,$$

then

$$P(A_1 A_2 \cdots A_n \mid B_1 B_2 \cdots B_m) = P(A_1 A_2 \cdots A_n).$$

Proof: In a straightforward manner we obtain

$$
\begin{aligned}
P(A_1 \cdots A_n \mid B_1 \cdots B_m) \\
&= P(A_1 \cdots A_n B_1 \cdots B_m)/P(B_1 \cdots B_m) \\
&= P(A_1) \cdots P(A_n) P(B_1) \cdots P(B_m)/P(B_1) \cdots P(B_m) \\
&= P(A_1 \cdots A_n).
\end{aligned}
$$

Proposition 4. If $\mathcal{C}$ is a class of independent events, and if each event in some subset of $\mathcal{C}$ is replaced by its complement, then the new class of events is also a class of independent events.

Proof: Let $A_1, \cdots, A_n, B_1, \cdots, B_m$ be any $m + n$ events in $\mathcal{C}$, m and n being arbitrary; we wish to prove that

$$P(A_1 \cdots A_n B_1^c \cdots B_m^c) = P(A_1) \cdots P(A_n) P(B_1^c) \cdots P(B_m^c).$$

We shall prove this by induction on m, for all n. For $m = 1$ the proof for all n is the same as the proof of Proposition 1. We now assume it to be true for some (fixed) m and all n; then for the case $m + 1$ we have, by induction hypothesis,

$$P(A_1 \cdots A_n B_1{}^c \cdots B_m{}^c B_{m+1}^c)$$

$$= P(A_1 \cdots A_n B_1{}^c \cdots B_m{}^c) - P(A_1 \cdots A_n B_{m+1} B_1{}^c \cdots B_m{}^c B_{m+1})$$

$$= P(A_1) \cdots P(A_n) P(B_1{}^c) \cdots P(B_m{}^c)(1 - P(B_{m+1}))$$

$$= \prod_{j=1}^{n} P(A_j) \prod_{k=1}^{m+1} P(B_k{}^c),$$

which proves the theorem.

The question arises whether the definition of a class of independent events given above is too strong. In other words, perhaps all that is needed is to stipulate "pairwise" independence—that is, that every *pair* of elements in ⑬ are independent, and that from this we might be able to deduce the statement given in the definition. The answer to this question is negative as is shown by the following example due to S. N. Bernstein. His example, which follows, shows that it is possible to have a collection of events that are not independent in the sense of the above definition and yet are pairwise independent. Suppose

$$\Omega = \{\omega_1, \omega_2, \omega_3, \omega_4\}, \qquad \text{where } P\{\omega_i\} = \tfrac{1}{4}, 1 \le i \le 4.$$

Let

$$A = \{\omega_1, \omega_2\}, \qquad B = \{\omega_1, \omega_3\}, \qquad C = \{\omega_1, \omega_4\},$$

and consider the class of events ⑬ $= \{A, B, C\}$. We observe that

$$P(A) = P(B) = P(C) = \tfrac{1}{2},$$

and

$$AB = AC = BC = \{\omega_1\}.$$

Thus,

$$P(AB) = P(AC) = P(BC) = \tfrac{1}{4} = P(A)P(B)$$

$$= P(A)P(C) = P(B)P(C).$$

However,

$$P(ABC) = P(\omega_1) = \tfrac{1}{4} \ne P(A)P(B)P(C) = \tfrac{1}{8}.$$

Thus, we need in our definition the requirement that the probability of joint occurrence of any finite number of events equals the product of their probabilities.

We also shall consider independence of classes of events.

Definition: Let $\{\mathcal{B}_\lambda, \lambda \in \Lambda\}$ be a family of classes of events. We say that these classes of events are independent if for every selection of events $B_\lambda \in \mathcal{B}_\lambda, \lambda \in \Lambda$, the events $\{B_\lambda, \lambda \in \Lambda\}$ are independent.

The simplest case of independent classes of events is given by the following proposition.

Proposition 5. Let $\mathcal{C} = \{B_\lambda, \lambda \in \Lambda\}$ be a class of independent events. For each $\lambda \in \Lambda$, let $\mathcal{B}_\lambda = \{B_\lambda, B_\lambda{}^c, \phi, \Omega\}$. Then $\{\mathcal{B}_\lambda, \lambda \in \Lambda\}$ is a family of independent classes of events.

Proof: This is merely a restatement of Proposition 4.

The classical practical example of independent events is the following: An urn contains N balls, among them being N_r red balls. A trial consists of selecting a ball at random from the urn, observing its color, and then returning it to the urn. Let A_n denote the event that a red ball is selected in the nth trial. It is very easy to verify that $\{A_1, A_2, \cdots\}$ are independent events.

EXERCISES

1. If A, B, and C are independent events, and if $P(C) > 0$, prove that $P(A \cup B \mid C) = P(A \cup B)$.
2. If $A_1, \cdots, A_n$ are events, $n \geq 2$, how many equations must be satisfied in order that these events be independent?
3. Let $(\Omega, \mathcal{C}, P)$ be the unit interval probability space (see Section 1.2), and let

$$A_n = \bigcup_{k=1}^{2^{n-1}} [(2k-1)/2^n, 2k/2^n], \qquad n = 1, 2, \cdots.$$

Prove that the events $\{A_n\}$ are independent.
4. Let Λ be any set, and let p be any function defined over Λ such that $0 \leq p(\lambda) \leq 1$ for all $\lambda \in \Lambda$. Prove that there exists a probability space $(\Omega, \mathcal{C}, P)$ and a class of events $\{B_\lambda, \lambda \in \Lambda\}$ in $\mathcal{C}$ which are independent and satisfy $P(B_\lambda) = p(\lambda)$ for all $\lambda \in \Lambda$.
5. Let $A_1, \cdots, A_n$ be n independent events, where

$$P(A_k) = p, \qquad 1 \leq k \leq n,$$

and where
$$0 < p < 1.$$

For every $\omega \in \Omega$, let $X(\omega)$ denote the number of these events that occur, that is, the number of these events that contain ω. (a) Prove that X is a random variable, and (b) find the distribution function and characteristic function of X.

3.2.   Independent Random Variables

In this section we define and characterize independence of random variables and deduce some special properties.

Definition: Let $\{X_\lambda, \lambda \in \Lambda\}$ be a family of random variables. They are said to be independent if for every positive integer n and every n distinct elements $\lambda_1, \cdots, \lambda_n$ in Λ, then

$$F_{X_{\lambda_1}, \cdots, X_{\lambda_n}}(x_1, \cdots, x_n) = \prod_{j=1}^{n} F_{X_{\lambda_j}}(x_j)$$

for all $\mathbf{x} \in E^{(n)}$.

It is easily seen that the random variables $\{X_\lambda, \lambda \in \Lambda\}$ are independent if and only if those of every finite subset are independent. Further, because of Lemma 1 of Section 2.2, the n random variables $X_1, \cdots, X_n$ are independent if and only if

$$F_{X_1, \cdots, X_n}(x_1, \cdots, x_n) = \prod_{j=1}^{n} F_{X_i}(x_j)$$

for every $\mathbf{x} \in E^{(n)}$.

The easiest example of a probability space over which independent random variables are defined is as follows: Let $\{F_\lambda, \lambda \in \Lambda\}$ be any set of distribution functions where the F_λ are not necessarily distinct for distinct λ. For every finite subset $\{\lambda_1, \cdots, \lambda_n\} \subset \Lambda$ we define

$$F_{\lambda_1, \cdots, \lambda_n}(x_1, \cdots, x_n) = \prod_{j=1}^{n} F_{\lambda_j}(x_j)$$

for all $\mathbf{x} \in E^{(n)}$. It is easy to verify (see Exercise 2 in Section 2.3) that the functions $\{F_{\lambda_1, \cdots, \lambda_n}\}$ just defined satisfy the hypotheses of the Daniell-Kolmogorov theorem (Theorem 2 in Section 2.3), and thus by that theorem there exists a probability space $(\Omega, \mathfrak{a}, P)$ and a set of random variables $\{X_\lambda, \lambda \in \Lambda\}$ defined over Ω such that

$$F_{\lambda_1, \cdots, \lambda_n}(x_1, \cdots, x_n) = P(\bigcap_{j=1}^{n} [X_{\lambda_j} \leq x_j]).$$

For each λ, however,

$$F_\lambda(x) = P[X_\lambda \leq x],$$

and thus by the way $F_{\lambda_1, \cdots, \lambda_n}$ is defined above, we have

$$P(\bigcap_{j=1}^{n} [X_{\lambda_j} \leq x_j]) = \prod_{j=1}^{n} P[X_{\lambda_j} \leq x_j],$$

thus satisfying the definition of independence.

Theorem 1. If $X_1, \cdots, X_n$ are independent random variables, if $f_1, \cdots, f_n$ are real- or complex-valued Borel-measurable functions, and if $Ef_j(X_j)$ exists for $1 \leq j \leq n$, then $E(\prod_{j=1}^{n} f_j(X_j))$ exists and

$$E(\prod_{j=1}^{n} f_j(X_j)) = \prod_{j=1}^{n} Ef_j(X_j).$$

Proof: Denote $G(x_1, \cdots, x_n) = \prod_{j=1}^{n} F_{X_j}(x_j)$. Using the easily verified fact that $\mu_G(\mathbf{a}, \mathbf{b}] = \prod_{j=1}^{n}(F_{X_j}(b_j) - F_{X_j}(a_j))$ (which implies that μ_G is a product measure), we may use Fubini's theorem to obtain

$$\prod_{j=1}^{n} E(f_j(X_j)) = \prod_{j=1}^{n} \int_{-\infty}^{\infty} f_j(x_j) \, dF_{X_j}(x_j)$$

$$= \int \cdots \int (\prod_{j=1}^{n} f_j(x_j)) \, dG(x_1, \cdots, x_n)$$

$$= E(\prod_{j=1}^{n} f_j(X_j)). \quad \text{Q.E.D.}$$

Corollary 1. If $X_1, \cdots, X_n$ are n independent random variables with finite second moments, then

$$\text{Var} \left(\sum_{k=1}^{n} X_k \right) = \sum_{k=1}^{n} \text{Var} (X_k).$$

Proof: By the definition of variance and by Theorem 1 one obtains

$$\text{Var} \left(\sum_{k=1}^{n} X_k \right) = E \left(\sum_{k=1}^{n} X_k - E \left(\sum_{k=1}^{n} X_k \right) \right)^2$$

$$= E \left(\sum_{k=1}^{n} (X_k - EX_k) \right)^2$$

$$= \sum_{k=1}^{n} E (X_k - EX_k)^2 + \sum_{j \neq k} E ((X_j - EX_j)(X_k - EX_k))$$

$$= \sum_{k=1}^{n} \text{Var} (X_k),$$

which proves the theorem.

One of the best criteria for determining independence is the following theorem.

Theorem 2. Let $X_1, \cdots, X_n$ be n random variables whose joint characteristic function is denoted by

$$f_{\mathbf{X}}(\mathbf{u}) = f_{X_1, \cdots, X_n}(u_1, \cdots, u_n)$$

and whose individual characteristic functions are

$$f_{X_j}(u_j), 1 \leq j \leq n.$$

Then $X_1, \cdots, X_n$ are independent if and only if

$$f_{\mathbf{X}}(\mathbf{u}) = \prod_{j=1}^{n} f_{X_j}(u_j) \qquad \text{for all} \quad \mathbf{u} \in E^{(n)}.$$

Proof: Suppose that $X_1, \cdots, X_n$ are independent. Since $\exp(iux)$ is continuous in x and is therefore a Borel-measurable function in x, we have by Theorem 1 that

$$f_{\mathbf{X}}(\mathbf{u}) = E\{\exp[i(\mathbf{u}, \mathbf{X})]\} = E[\prod_{j=1}^{n} \exp(iu_jX_j)]$$

$$= \prod_{j=1}^{n} E[\exp(iu_jX_j)] = \prod_{j=1}^{n} f_{X_j}(u_j),$$

which proves the condition is necessary. Conversely, if the condition is true, then by the remark made in the proof of Theorem 1,

$$\int \cdots \int \exp[i(\mathbf{u}, \mathbf{x})] \, dF_{\mathbf{X}}(\mathbf{x}) = \prod_{j=1}^{n} \int \exp(iu_jx_j) \, dF_{X_j}(x_j)$$

$$= \int \cdots \int \exp[i(\mathbf{u}, \mathbf{x})] \, d(\prod_{j=1}^{n} F_{X_j}(x_j)).$$

The independence of $X_1, \cdots, X_n$ follows from the uniqueness theorem (Theorem 1 in Section 2.5) and the definition of independence, which concludes the proof of the theorem.

A most frequently used result is the following corollary.

Corollary 2. If $X_1, \cdots, X_n$ are independent random variables with characteristic functions $f_1, \cdots, f_n$ respectively, and if f is the characteristic function of $\sum_{j=1}^{n} X_j$, then

$$f(u) = \prod_{j=1}^{n} f_j(u).$$

Proof: Since $X_1, \cdots, X_n$ are independent, then

$$f_{X_1,\cdots,X_n}(u_1, \cdots, u_n) = \prod_{j=1}^{n} f_j(u_j)$$

by Theorem 2. Now let

$$u_1 = u_2 = \cdots = u_n = u,$$

and we obtain the conclusion of the corollary.

The true significance of convolution which we defined earlier is seen in the following corollary.

Corollary 3. If X and Y are independent random variables, and if F_{X+Y}, F_X, and F_Y are the distribution functions of $X + Y$, X, and Y, respectively, then

$$F_{X+Y} = F_X * F_Y.$$

Proof: If f_{X+Y}, f_X, and f_Y denote the characteristic functions of $X + Y$, X, and Y respectively, then by Corollary 2,

$$f_{X+Y}(u) = f_X(u) f_Y(u).$$

By Theorem 2 of Section 2.5 $f_X(u) f_Y(u)$ is the characteristic function of $F_X * F_Y$. Hence by the uniqueness theorem,

$$F_{X+Y} = F_X * F_Y$$

which concludes the proof.

Theorem 3. If $\{X_\lambda, \lambda \in \Lambda\}$ are independent random variables, and if $\{f_\lambda, \lambda \in \Lambda\}$ is a corresponding set of Borel-measurable functions, then $\{f_\lambda(X_\lambda), \lambda \in \Lambda\}$ are independent random variables.

Proof: Let $\{\lambda_1, \cdots, \lambda_n\}$ be a subset of n distinct elements of Λ. We need only prove that $f_{\lambda_1}(X_{\lambda_1}), \cdots, f_{\lambda_n}(X_{\lambda_n})$ are independent for every such finite subset of Λ. The joint characteristic function of these random variables is

$$f(u_1, \cdots, u_n) = E \exp \left(i \sum_{j=1}^{n} u_j f_{\lambda_j}(X_{\lambda_j}) \right)$$

$$= \int \cdots \int \exp \left\{ i \sum_{j=1}^{n} u_j f_{\lambda_j}(x_j) \right\} dF_{X_{\lambda_1}, \cdots, X_{\lambda_n}}(x_1, \cdots, x_n)$$

$$= \int \cdots \int \exp \left\{ i \sum_{j=1}^{n} u_j f_{\lambda_j}(x_j) \right\} d\left\{ \prod_{j=1}^{n} F_{X_{\lambda_j}}(x_j) \right\}$$

$$= \prod_{j=1}^{n} \int \exp \left\{ i u_j f_{\lambda_j}(x_j) \right\} dF_{X_{\lambda_j}}(x_j) \quad \text{(by Exercise 10)}$$

$$= \prod_{j=1}^{n} E \exp \left(i u_j f_{\lambda_j}(X_{\lambda_j}) \right),$$

which by Theorem 2 proves that $\{f_{\lambda_1}(X_1), \cdots, f_{\lambda_n}(X_n)\}$ are independent.

Corollary. The random variables $\{X_\lambda, \lambda \in \Lambda\}$ are independent if and only if the classes of events $\{\sigma\{X_\lambda\}, \lambda \in \Lambda\}$ are independent.

Proof: The "if" part is immediate, since for all real x,

$$[X_\lambda \leq x] \in \sigma\{X_\lambda\}$$

for all λ. In order to prove the "only if" part, we assume that $\{X_\lambda, \lambda \in \Lambda\}$ are independent and let $\{B_\lambda, \lambda \in \Lambda\}$ be any corresponding family of Borel sets of real numbers. Let χ_{B_λ} denote the indicator of the set B_λ. Now $\{\chi_{B_\lambda}, \lambda \in \Lambda\}$ are Borel-measurable functions. Further, it is easy to verify that

$$\chi_{B_\lambda}(X_\lambda) = I_{[X_\lambda \in B_\lambda]}.$$

By Theorem 3, $\{\chi_{B_\lambda}(X_\lambda), \lambda \in \Lambda\}$ are independent random variables. This implies that $\{[X_\lambda \in B_\lambda], \lambda \in \Lambda\}$ are independent events, and hence $\{\sigma\{X_\lambda\}, \lambda \in \Lambda\}$ are independent classes of events.

Characteristic functions become quite useful when one wishes to find the distribution function of a sum of independent random variables. Several important examples of this follow:

(a) Let X and Y be independent random variables whose distributions are $\mathfrak{N}(\mu_1, \sigma_1^2)$ and $\mathfrak{N}(\mu_2, \sigma_2^2)$, respectively. The problem is to determine the distribution function of $X + Y$. From Section 2.5 we see that the characteristic functions of X and Y are

$$f_X(u) = \exp\left\{iu\mu_1 - \frac{\sigma_1^2 u^2}{2}\right\}$$

and

$$f_Y(u) = \exp\left\{iu\mu_2 - \frac{\sigma_2^2 u^2}{2}\right\}.$$

By Corollary 2 we have that the characteristic function of $X + Y$ is

$$f_{X+Y}(u) = f_X(u)f_Y(u)$$
$$= \exp\left\{iu(\mu_1 + \mu_2) - (\sigma_1^2 + \sigma_2^2)u^2/2\right\}.$$

Hence by the uniqueness theorem the distribution of $X + Y$ is $\mathfrak{N}(\mu_1 + \mu_2, \sigma_1^2 + \sigma_2^2)$.

(b) Let X and Y be independent random variables whose distributions are $\mathcal{P}(\lambda_1)$ and $\mathcal{P}(\lambda_2)$, respectively. From Section 2.5 we obtain

$$f_X(u) = \exp\left\{\lambda_1(e^{iu} - 1)\right\}$$

and
$$f_Y(u) = \exp\{\lambda_2(e^{iu} - 1)\}.$$

By Corollary 2, we have

$$f_{X+Y}(u) = f_X(u)f_Y(u) = \exp\{(\lambda_1 + \lambda_2)(e^{iu} - 1)\}.$$

Hence by the uniqueness theorem the distribution function of $X + Y$ is $\mathcal{P}(\lambda_1 + \lambda_2)$.

(c) Suppose X and Y are independent random variables whose distributions are $B(n_1, p)$ and $B(n_2, p)$, respectively. From Section 2.5 we obtain

$$f_X(u) = ((1 - p) + pe^{iu})^{n_1}$$

and

$$f_Y(u) = ((1 - p) + pe^{iu})^{n_2},$$

and thus

$$f_{X+Y}(u) = ((1 - p) + pe^{iu})^{n_1+n_2},$$

that is, the distribution of $X + Y$ is $B(n_1 + n_2, p)$.

A useful tool for studying sums of independent random variables is the concentration function that we consider briefly.

Definition: If F is a distribution function, its concentration function Q_F is defined on $[0, \infty)$ by

$$Q_F(L) = \sup \{F(x + L) - F(x - 0) \mid -\infty < x < \infty\},$$

$$0 \le L < \infty.$$

If X is a random variable, its concentration function Q_X is defined by

$$Q_X(L) = \sup \{P[x \le X \le x + L] \mid -\infty < x < \infty\}$$

for all $L \ge 0$.

The concentration function tells us how concentrated the probability measure is on the line.

Theorem 4. If X and Y are independent random variables, then $Q_X(L) \ge Q_{X+Y}(L)$ for all $L \ge 0$.

Proof: By the definition of Q and by Corollary 3 to Theorem 2 we have

$$Q_{X+Y}(L) = \sup \{P[x \leq X + Y \leq x + L] \mid -\infty < x < \infty\}$$

$$= \sup_x \left\{ \int F_X(x + L - y) \, dF_Y(y) \right.$$

$$\left. - \int F_X(x - 0 - y) \, dF_Y(y) \right\}$$

$$= \sup_x \int (F_X(x + L - y) - F_X(x - 0 - y)) \, dF_Y(y)$$

$$\leq \sup_x \int Q_X(L) \, dF_Y(y) = Q_X(L),$$

which proves the theorem.

The intuitive content of Theorem 4 is that if two independent random variables are added, the spread of probability mass of their distribution extends over a wider range.

EXERCISES

1. Prove the following theorem: If $\{X_n\}$ is a sequence of independent random variables, if $0 = n_0 < n_1 < n_2 < \cdots$ is an increasing sequence of integers, and if f_k is a Borel-measurable function whose domain is $E^{(n_k - n_{k-1})}$, then

$$\{ f_k(X_{n_{k-1}+1}, \cdots, X_{n_k}), \ k = 1, 2, \cdots \}$$

are independent random variables and

$$\{\sigma\{X_{n_{k-1}+1}, \cdots, X_{n_k}\}, \qquad k = 1, 2, \cdots\}$$

are independent classes of events.

2. For each $x \in [0, 1]$ define $R_n(x)$ by $R_n(x) = 1$ if the nth digit in the dyadic expansion of x is 1, and $R_n(x) = -1$ if the nth digit in the dyadic expansion of x is 0. (Do not allow finitely many zeros in any such expression.) The functions $\{R_n\}$ are called the *Rademacher functions* Let $(\Omega, \mathcal{C}, P)$ be the unit interval probability space. Prove

that $\{R_n\}$ are independent random variables with the same distribution function.

3. Let $X_1, \cdots, X_n$ be n independent random variables, all with the same Cauchy distribution. Prove that X_1 and $(X_1 + \cdots + X_n)/n$ have the same distributions.

4. Let F be a distribution function with concentration function Q_F. Prove: (a) there exists an x_L such that

$$Q_F(L) = F(x_L + L) - F(x_L - 0),$$

and (b) F is continuous if and only if $Q_F(0) = 0$.

5. Let $\{A_n\}$ be a sequence of events. Prove that $\{A_n\}$ are independent events if and only if the random variables $\{I_{A_n}\}$ are independent.

6. Let $\{X_n\}$ be a sequence of random variables. Prove that $\{X_n\}$ are independent if and only if

$$F_{X_1,\cdots,X_n}(x_1, \cdots, x_n) = \prod_{j=1}^{n} F_{X_j}(x_j)$$

for all n and all $(x_1, \cdots, x_n) \in E^{(n)}$.

7. The purpose of this exercise is to show that if $\{F_n\}$ is any sequence of distribution functions, then there exists a sequence of independent random variables $\{X_n\}$ defined over the unit interval probability space such that the distribution function of X_n is F_n for all n.

(a) Let $\{R_n\}$ be the Rademacher functions, and let

$$Y_1 = R_1/2 + R_3/2^2 + R_6/2^3 + R_{10}/2^4 + R_{15}/2^5 + \cdots$$

$$Y_2 = R_2/2 + R_4/2^2 + R_7/2^3 + R_{11}/2^4 + R_{16}/2^5 + \cdots$$

$$Y_3 = R_5/2 + R_8/2^2 + R_{12}/2^3 + \cdots, \cdots$$

Prove that $\{Y_n\}$ are independent random variables, each uniformly distributed over $[-1, 1]$, that is, each with density $\varphi(x) = \frac{1}{2}$ if $x \in [-1, 1]$ and $\varphi(x) = 0$ if $|x| > 1$.

(b) Find a Borel-measurable function φ_n such that the distribution function of $\varphi_n(Y_n)$ is F_n for every n. If

$$X_n = \varphi_n(Y_n),$$

then the point made at the beginning of this exercise is established for $\{X_n\}$.

8. Let $X_1, \cdots, X_n$ be n independent random variables where

$$P[X_k = 1] = p \quad \text{and} \quad P[X_k = 0] = 1 - p \quad \text{for} \quad 1 \leq k \leq n,$$

and where $0 < p < 1$. Let $X = X_1 + \cdots + X_n$. Find the distribution function and characteristic function of X.

9. *Definition*: A collection of vector random variables $\{\mathbf{X}_\lambda, \ \lambda \in \Lambda\}$ which do not necessarily have the same dimension are said to be independent if the sigma fields $\{\sigma\{\mathbf{X}_\lambda\}, \ \lambda \in \Lambda\}$ are independent. Prove that the random vectors $\mathbf{X}_1, \cdots, \mathbf{X}_n$ are independent if and only if their joint characteristic function is equal to the product of the individual characteristic functions, that is,

$$E \exp\left(i \sum_{j=1}^{n} (\mathbf{u}_j, \mathbf{X}_j)\right) = \prod_{j=1}^{n} E \exp\left(i(\mathbf{u}_j, \mathbf{X}_j)\right),$$

where $\mathbf{u}_j$ is a real vector whose dimension is that of $\mathbf{X}_j$.

10. Prove the remark in parenthesis in the proof of Theorem 1.

3.3. The Zero-One Law

The assumption of independence is a very strong one, and therefore it should yield some very strong results. One of the strongest results obtained from independence is a theorem due to A. N. Kolmogorov, referred to as the "zero-one law." In this section we prove the zero-one law and the Borel lemmas. We begin with one of the most frequently referred to results in probability theory.

Theorem 1 (The "Borel-Cantelli Lemma"). If $\{A_n\}$ is a sequence of events, and if $\sum_{n=1}^{\infty} P(A_n) < \infty$, then the probability that infinitely many of these events occur is zero. (Notationally: $P[A_n \text{ i.o.}] = 0$, where "i.o." denotes "infinitely often.")

Proof: The event $[A_n \text{ i.o.}]$ means that for every n there is a $k \geq n$ such that A_k occurs; that is,

$$[A_n \text{ i.o.}] = \bigcap_{m=1}^{\infty} \bigcup_{k=m}^{\infty} A_k = \limsup_{n \to \infty} A_n.$$

Now by Boole's inequality we have for every m

$$P[A_n \text{ i.o.}] \leq P\left(\bigcup_{k=m}^{\infty} A_k\right) \leq \sum_{k=m}^{\infty} P(A_k) \to 0$$

as $m \to \infty$ since, by hypothesis, $\sum P(A_n) < \infty$. Thus $P[A_n \text{ i.o.}] = 0$. Q.E.D.

The converse of this theorem is not necessarily true. If (Ω, α, P) is the unit interval probability space, and if $A_n = (0, 1/n)$, then every point $x \in [0, 1]$ is in a finite number of A_n, so

$$[A_n \text{ i.o.}] = \phi \qquad \text{and} \qquad P[A_n \text{ i.o.}] = 0.$$

However, $\sum P(A_n) = \sum 1/n$ is not a convergent series.

One of the more usual uses of the Borel-Cantelli lemma is that of truncation. Let $\{X_n\}$ be a sequence of random variables, and let $\{c_n\}$ be a sequence of positive real numbers. The truncation of X_n at c_n is defined to be the random variable

$$Y_n = X_n I_{[|X_n| \le c_n]}.$$

The following two corollaries illustrate how truncation will be used.

Corollary 1. If

$$\sum_{n=1}^{\infty} P[\,|X_n| > c_n\,] < \infty,$$

and if

$$P[Y_n \to X \qquad \text{as } n \to \infty] = 1,$$

then

$$P[X_n \to X \qquad \text{as } n \to \infty] = 1.$$

Proof: Let

$$A = [Y_n \to X],$$

let

$$B_n = [\,|X_n| > c_n\,],$$

let

$$B = \limsup B_n = [B_n \text{ i.o.}],$$

and let

$$C = [X_n \to X].$$

By the Borel-Cantelli lemma, $P(B) = 0$. Since B is the event that $X_n \ne Y_n$ for infinitely many values of n, then B^c is the event that $X_n = Y_n$ for all but a finite number of values for n. But $P(B^c) = 1$ and by hypothesis $P(A) = 1$, so $P(AB^c) = 1$. However, $C \supset AB^c$, and hence $P(C) = 1$. Q.E.D.

Corollary 2. If $\{X_n\}$ is a sequence of random variables, if $\{c_n\}$ is a sequence of positive numbers, and if $\sum P[\,|X_n| \geq c_n] < \infty$, then

$$P[\textstyle\sum X_n \text{ converges}] = P[\textstyle\sum X_n I_{[\,|X_n|<c_n]} \text{ converges}].$$

Proof: The proof of this corollary is step by step the same as the proof of Corollary 1.

Although, in general, the converse of the Borel-Cantelli lemma is not true, it is true under the assumption of independence. We thus obtain our first zero-one law as follows.

Theorem 2 (Borel Lemmas). If $\{A_n\}$ is a sequence of independent events, then

(a) $P[A_n \text{ i.o.}] = 0$ if and only if $\sum P(A_n) < \infty$, and

(b) $P[A_n \text{ i.o.}] = 1$ if and only if $\sum P(A_n) = \infty$.

Proof: Because of Theorem 1 we need only prove that $\sum P(A_n) = \infty$ implies that $P[A_n \text{ i.o.}] = 1$. Since

$$[A_n \text{ i.o.}] = \bigcap_{m=1}^{\infty} \bigcup_{k=m}^{\infty} A_k,$$

we have

$$P[A_n \text{ i.o.}] = \lim_{m \to \infty} P(\bigcup_{k=m}^{\infty} A_k),$$

or

$$1 - P[A_n \text{ i.o.}] = \lim_{m \to \infty} P(\bigcap_{k=m}^{\infty} A_k^c).$$

Since $\{A_n^c\}$ are also independent (see Proposition 4 of Section 3.1), we have

$$1 - P[A_n \text{ i.o.}] = \lim_{m \to \infty} \lim_{n \to \infty} P(\bigcap_{k=m}^{n} A_k^c)$$

$$= \lim_{m \to \infty} \lim_{n \to \infty} \prod_{k=m}^{n} P(A_k^c)$$

$$= \lim_{m \to \infty} \prod_{k=m}^{\infty} (1 - P(A_k)).$$

Now the hypothesis $\sum P(A_n) = \infty$ implies that the infinite product $\prod_{k=m}^{\infty}(1 - P(A_k))$ diverges to zero for all m, which yields $P[A_n \text{ i.o.}] = 1$. Q.E.D.

Notation: If A and B are events, we denote

$$A \triangle B = AB^c \cup A^cB.$$

This is called the *symmetric difference* of A and B. Clearly $A \triangle B \in \mathcal{C}$.

Lemma 1. If $\{A_n\}$ is a sequence of events, if A is an event, and if $P(A_n \triangle A) \to 0$, then $P(A_n) \to P(A)$.

Proof: We observe that

$$0 \le |P(A_n) - P(A)| = \left| \int (I_{A_n} - I_A)\, dP \right|$$

$$\le \int |I_{A_n} - I_A|\, dP = P(A_n \triangle A),$$

which yields the conclusion.

Lemma 2. Let $\{X_n\}$ be a sequence of random variables. For every $B \in \sigma\{X_1, X_2, \cdots\}$ and for every $\epsilon > 0$ there exist a finite set of integers $\{n_1, \cdots, n_k\}$ and an event $B_\epsilon \in \sigma\{X_{n_1}, \cdots, X_{n_k}\}$ such that $P(B \triangle B_\epsilon) < \epsilon$.

Proof: Consider for every finite subset $\{m_1, \cdots, m_j\} \subset \{1, 2, \cdots\}$ the sigma field $\sigma\{X_{m_1}, \cdots, X_{m_j}\}$, and let $\mathfrak{F}$ denote the union of all such sigma fields. It is easily verified that $\mathfrak{F}$ is a field of events and that $\sigma\{X_1, X_2, \cdots\}$ is the smallest sigma field containing $\mathfrak{F}$. It is a standard result in measure theory that for every $\epsilon > 0$ there is a $B_\epsilon \in \mathfrak{F}$ such that $P(B \triangle B_\epsilon) < \epsilon$. But $B_\epsilon \in \mathfrak{F}$ implies that there exists a finite set of integers $\{n_1, \cdots, n_k\}$ such that $B_\epsilon \in \sigma\{X_{n_1}, \cdots, X_{n_k}\}$. Q.E.D.

Definition: Let $\{X_n\}$ be a sequence of random variables. The tail sigma field of this sequence is defined to be

$$\mathfrak{I} = \bigcap_{n=1}^{\infty} \sigma\{X_n, X_{n+1}, \cdots\}.$$

We next prove the main theorem of this section.

Theorem 3 (Kolmogorov's Zero-or-One Law). If $\{X_n\}$ is a sequence of independent random variables, and if $B \in \Im$, the tail sigma field of $\{X_n\}$, then $P(B) = 0$ or 1.

Proof: Let $\epsilon_n > 0$ be such that $\epsilon_n \to 0$ as $n \to \infty$. Since

$$\Im \subset \sigma\{X_1, X_2, \cdots\},$$

then by Lemma 2 there exist a finite set of integers $\{n_{1,1}, \cdots, n_{1,k_1}\}$ and an event

$$B_1 \in \sigma\{X_{n_{1,1}}, \cdots, X_{n_{1,k_1}}\}$$

such that $P(B \bigtriangleup B_1) < \epsilon_1$. (We assume

$$n_{j,1} < n_{j,2} < \cdots < n_{j,k_j} \qquad \text{for} \quad j = 1, 2, \cdots$$

in this proof.) Since

$$B \in \Im \subset \sigma\{X_{n_{1,k_1}+1}, \cdots\},$$

there exist a finite set of integers $\{n_{2,1}, \cdots, n_{2,k_2}\}$, all greater than n_{1,k_1}, and an event

$$B_2 \in \sigma\{X_{n_{2,1}}, \cdots, X_{n_{2,k_2}}\}$$

such that $P(B \bigtriangleup B_2) < \epsilon_2$. In general, then, there exists an increasing sequence of integers $\{m_1, m_2, \cdots\}$ and a sequence of events $\{B_k\}$ such that

$$B_k \in \sigma\{X_{m_k}, \cdots, X_{m_{k+1}-1}\}$$

and such that $P(B \bigtriangleup B_k) < \epsilon_k$. By the theorem in Exercise 1 in Section 3.2, the events $\{B_k\}$ are independent. Now consider the sequence of events $\{B_n \cap B_{n+1}\}$. We have

$$P(B \bigtriangleup (B_n B_{n+1}))$$
$$= P(B \cap (B_n B_{n+1})^c) + P(B^c \cap B_n B_{n+1})$$
$$\leq P(B(B_n{}^c \cup B_{n+1}^c)) + P(B^c B_n) + P(B^c B_{n+1})$$
$$\leq P(BB_n{}^c) + P(B^c B_n) + P(BB_{n+1}^c) + P(B^c B_{n+1})$$
$$= P(B \bigtriangleup B_n) + P(B \bigtriangleup B_{n+1}) \to 0 \qquad \text{as} \quad n \to \infty.$$

Thus, by Lemma 1,

$$P(B_n B_{n+1}) \to P(B).$$

But because of independence of $\{B_n\}$,

$$P(B_n B_{n+1}) = P(B_n)P(B_{n+1}) \to P^2(B) \qquad \text{as} \quad n \to \infty.$$

Hence

$$P(B) = P^2(B) \quad \text{or} \quad P(B) = 0 \quad \text{or} \quad 1. \quad \text{Q.E.D.}$$

The Kolmogorov zero-or-one law will be applied many times in the future. At this point we illustrate its uses and meaning.

Corollary 1. If $\{X_n\}$ is a sequence of independent random variables with tail sigma field $\mathfrak{J}$, and if Y is a random variable such that $\sigma\{Y\} \subset \mathfrak{J}$, then Y is constant over an event of probability one.

Proof: Let F be the distribution function of Y. There exists a y such that $F(y) > 0$. By the Kolmogorov zero-or-one law, $F(y) = 1$. Let

$$y_0 = \inf \{y \mid F(y) = 1\}.$$

Clearly y_0 is finite, for otherwise $F(x) \equiv 1$, which is impossible. It is clear that $y < y_0$ implies $F(y) = 0$. Thus $P[Y = y_0] = 1$, which concludes the proof.

Corollary 2. If $\{X_n\}$ is a sequence of independent random variables, then

$$P[\sum_{n=1}^{\infty} X_n \text{ converges}] = 0 \text{ or } 1.$$

Proof: It is easy to see that if $C = [\sum_{n=1}^{\infty} X_n \text{ converges}]$, then $C = [\sum_{n=N}^{\infty} X_n \text{ converges}]$ for every N. Hence $C \in \sigma\{X_N, X_{N+1}, \cdots\}$ for all N; that is, $C \in \mathfrak{J}$. Hence by the zero-or-one law, $P(C) = 0$ or 1. Q.E.D.

Let $\{X_n\}$ be a sequence of random variables, and let us consider a power series with random coefficients: $\sum_{n=1}^{\infty} X_n z^n$. The radius of convergence of this series,

$$R = 1/\limsup_{n \to \infty} \sqrt[n]{|X_n|},$$

is a random variable that can possibly equal infinity with positive probability.

Corollary 3. If $\{X_n\}$ is a sequence of independent random variables, then the radius of convergence of the power series $\sum X_n z^n$ is a constant (possibly infinite) over an event of probability one.

Proof: If R is as above, then also

$$R = 1/\limsup_{n \to \infty} {}^{n+k}\!\sqrt{|X_{n+k}|}$$

for every k. Hence R is measurable with respect to $\sigma\{X_k, X_{k+1}, \cdots\}$ for all k; that is, $\sigma\{R\} \subset \mathfrak{I}$. By Corollary 1, R is constant over an event of probability one.

EXERCISES

1. Let $\{X_n\}$ be a sequence of random variables.

   (a) Prove that $[X_1 + X_2 \leq z] \in \sigma\{X_1, X_2\}$. (Hint: First prove that

   $$[X_1 + X_2 < z] = \cup_r [X_1 < r][X_2 < z - r],$$

   where the union is taken over all rational numbers.)

   (b) Prove that

   $$\left[\sum_{n=N}^{\infty} X_n \text{ converges}\right] \in \sigma\{X_N, X_{N+1}, \cdots\}.$$

   (Hint: First verify that

   $$\left[\sum_{n=N}^{\infty} X_n \text{ converges}\right]$$

   $$= \bigcap_{n=1}^{\infty} \bigcup_{m=N}^{\infty} \bigcap_{k=m}^{\infty} \bigcap_{l=k}^{\infty} [-1/n < X_k + \cdots + X_l < 1/n].)$$

2. (*True essence of zero-or-one-ism.*) Let $\{X_\lambda, \lambda \in \Lambda\}$ be an infinite set of independent random variables. Let $\mathfrak{F}$ be the set of all finite subsets of Λ, and let

   $$\mathfrak{I} = \bigcap_{\Phi \in \mathfrak{F}} \sigma\{X_\lambda \mid \lambda \notin \Phi\}.$$

   Prove that $\mathfrak{I}$ is a sigma field and that if $B \in \mathfrak{I}$, then $P(B) = 0$ or 1.

3. Let $(\Omega, \mathfrak{a}, P)$ be the unit interval probability space. For $n = 1, 2, \cdots$, define

   $$X_n = 2I_{[0,1/2+(1/2)n]} + 3I_{[1/2+(1/2)n,1]}.$$

   (a) What is the tail sigma field of $\{X_n\}$?

   (b) Are the random variables $\{X_n\}$ independent?

CHAPTER 4

Basic Limiting Operations

4.1. Convergence of Distribution Functions

This chapter deals with the last of the basic analytical background needed for proving the fundamental limit theorems of probability theory. The four basic types of convergence are: almost sure convergence, convergence in quadratic mean, convergence in probability, and convergence of distribution functions. This section is devoted to convergence of distribution functions and yields the Helly and Helly-Bray theorems.

First we reestablish some notation. If F is a function defined over k-dimensional Euclidean space $E^{(k)}$, and if F is not necessarily a distribution function, then μ_F will be a function over the set of all cells in $E^{(k)}$ as defined in Section 2.2. If, however, F has all the properties of a distribution function except perhaps that the total variation is not one, then μ_F will denote the Lebesgue-Stieltjes measure over $E^{(k)}$ determined by F. We shall in addition define the *closed cell* $[\mathbf{a}, \mathbf{b}]$ by

$$[\mathbf{a}, \mathbf{b}] = \underset{i=1}{\overset{k}{\times}} [a_i, b_i],$$

where $\mathbf{a} = (a_1, \cdots, a_k)$, $\mathbf{b} = (b_1, \cdots, b_k)$, and $a_i \leq b_i$, $1 \leq i \leq k$.

Notation: If F is a function defined over $E^{(k)}$, then *Cont F* will denote the set of points in $E^{(k)}$ at which F is continuous.

Definition: Let $\{F_n\}$ be a sequence of distribution functions over $E^{(k)}$. We say $\{F_n\}$ converges in the wide sense (Notation: $F_n \xrightarrow{w}$) if there exists a function F over $E^{(k)}$ such that $F_n(\mathbf{x}) \to F(\mathbf{x})$ as $n \to \infty$ at all $\mathbf{x} \in Cont\, F$, and such that

(c) F is continuous from the right in each coordinate, and

(d) $\mu_F(\mathbf{a}, \mathbf{b}] \geq 0$ for every cell $(\mathbf{a}, \mathbf{b}] \subset E^{(k)}$. In this case we write $F_n \xrightarrow{w} F$.

In the definition just given, (c) and (d) are the same conditions (c) and (d) in Theorem 1 of Section 2.2. By omitting requirements (a) and (b), it is possible that the total variation of F can be less than one. As an extreme case, consider the following: let $k = 1$, and let $F_n(x)$ be defined by

$$
F_n(x) = \begin{cases} 1 & \text{if } x \geq n \\[2mm] \tfrac{1}{2} & \text{if } -n \leq x < n \\[2mm] 0 & \text{if } x < -n. \end{cases}
$$

In this case $F_n \xrightarrow{w} F$, where $F(x) = \tfrac{1}{2}$ for all x.

Lemma 1. If $\{F_n\}$ is a sequence of distribution functions over $E^{(k)}$, and if $F_n \xrightarrow{w} F$, then F is continuous everywhere over $E^{(k)}$ except over a countable union of hyperplanes of the form $x_i = c$.

Proof: It should be observed that in the proof of Theorem 2 in Section 2.2, conditions (a) and (b) of that theorem were used only to establish $\mu_F(E^{(k)}) = 1$. In the definition of $F_n \xrightarrow{w} F$ given above, F does not necessarily satisfy (a) and (b), and hence it is possible for $\mu_F(E^{(k)}) < 1$. The proof of Theorem 3 of Section 2.2 does not depend at all on the fact that F determines a normed measure, and the same proof of that theorem goes through word for word here except that P is replaced by μ_F and $[X_i \leq t]$ is replaced by $\{\mathbf{x} \mid x_i \leq t\}$. Thus the lemma is true.

Lemma 2. If F is a distribution function over $E^{(k)}$, or if F is a limit in the wide sense of a sequence of distribution functions $\{F_n\}$ over $E^{(k)}$, then

$$
\mu_F\{\mathbf{x} \in E^{(k)} \mid x_j = c\} = 0
$$

if $x_j = c$ is not a hyperplane in the countable union of hyperplanes over which F is not continuous.

Proof: Let

$$H = \{\mathbf{x} \in E^{(k)} \mid x_j = c\}.$$

We may assume that F is a joint distribution function of k random variables. If we denote

$$H_n = \{\mathbf{x} \in E^{(k)} \mid -n < x_i < n \quad \text{for} \quad i \neq j, c - \epsilon < x_j \leq c\},$$

then

$$H_n \to \{\mathbf{x} \in E^{(k)} \mid c - \epsilon < x_j \leq c\},$$

and by the definition of μ_F, we have

$$\mu_F(H_n) \to F_j(c) - F_j(c - \epsilon),$$

where F_j is in the notation used in the proof of Theorem 3 in Section 2.2. Since by hypothesis, F_j is continuous at c, and since

$$\{\mathbf{x} \in E^{(k)} \mid c - \epsilon < x_j \leq c\} \to \{\mathbf{x} \in E^{(k)} \mid x_j = c\} \quad \text{as} \quad \epsilon \to 0 \; (\epsilon > 0),$$

we obtain the conclusion.

Theorem 1. Let $\{F_n\}$ be a sequence of distribution functions over $E^{(k)}$. If $\{F_n\}$ converges pointwise over a dense subset of $E^{(k)}$, then there exists one and only one function F defined over $E^{(k)}$ such that $F_n \xrightarrow{w} F$.

Proof: Let D denote the dense subset of $E^{(k)}$ over which $\{F_n\}$ converges pointwise. For every

$$\mathbf{x} = (x_1, \cdots, x_k) \in E^{(k)},$$

whether it is in D or not, let us define

$$F(\mathbf{x}) = \inf \{\lim_{n \to \infty} F_n(\mathbf{y}) \mid \mathbf{y} = (y_1, \cdots, y_k) \in D, x_i < y_i, 1 \leq i \leq k\}.$$

(Note: if $\mathbf{x} \in D$ it might occur that $F(\mathbf{x}) > \lim_{n \to \infty} F_n(\mathbf{x})$; for example, if $k = 1$ and $F_n = \chi_{[1/n, \infty)}$, then $F(0) = 1$ but $\lim_{n \to \infty} F_n(0) = 0$.) Clearly, $F(x_1, \cdots, x_k)$ is nondecreasing in each argument. We now show that F is continuous from the right in each argument. We do this for x_1. Let $\epsilon > 0$; we shall show that there exists a $\delta > 0$ such that if

$0 < t < \delta$, then

$$F(x_1 + t, x_2 \cdots, x_k) - F(x_1, \cdots, x_n) < \epsilon.$$

By the definition of F above, we know that there is a $(y_1, \cdots, y_k) \in D$ such that $y_j > x_j$ for $1 \leq j \leq k$ and such that

$$\lim_{n \to \infty} F_n(\mathbf{y}) < F(\mathbf{x}) + \tfrac{1}{2}\epsilon.$$

Let

$$0 < \delta < \min \{y_i - x_i, 1 \leq i \leq k\}.$$

By the monotonicity of F we may obtain that for every

$$\mathbf{w} \in D \cap \{\mathbf{z} \mid x_i < z_i < x_i + \delta\}, \qquad 0 \leq \lim_{n \to \infty} F_n(\mathbf{w}) - F(\mathbf{x}) < \tfrac{1}{2}\epsilon.$$

Again by the definition of F, if $0 < t < \delta$, and if $0 < r \leq \delta - t$, there is a point

$$\mathbf{w}^* = (w_1{}^*, \cdots, w_k{}^*) \in D \cap \{\mathbf{z} \mid x_i < z_i < x_i + r, 1 \leq i \leq k\}$$

such that

$$x_1 + t < w_1{}^* < x_1 + r + t$$

and which satisfies

$$0 \leq \lim_{n \to \infty} F_n(\mathbf{w}^*) - F(x_1 + t, x_2, \cdots, x_k) < \tfrac{1}{2}\epsilon.$$

Hence for $0 < t < \delta$,

$$0 \leq | F(x_1 + t, x_2, \cdots, x_k) - F(x_1, \cdots, x_k) |$$

$$\leq | F(x_1 + t, x_2, \cdots, x_k) - \lim_{n \to \infty} F_n(\mathbf{w}^*) |$$

$$+ | \lim_{n \to \infty} F_n(\mathbf{w}^*) - F(\mathbf{x}) | < \epsilon,$$

which concludes the proof that F is continuous from the right in each argument. (This same proof shows that F is continuous from above in all arguments.) We next show that $F_n(\mathbf{x}) \to F(\mathbf{x})$ as $n \to \infty$ for all $x \in Cont\ F$. Let $\mathbf{x} \in Cont\ F$ and let $\epsilon > 0$. Then there exists a $\delta > 0$ such that if $| \mathbf{y} - \mathbf{x} | < \delta$, then $| F(\mathbf{x}) - F(\mathbf{y}) | < \epsilon$. Let $(\mathbf{a}, \mathbf{b})$ be a cell in $E^{(k)}$ that contains $\mathbf{x}$, has diameter less than δ, and is such that $\mathbf{a} \in D$, $\mathbf{b} \in D$. Since by the definition of F,

$$\lim_{n \to \infty} F_n(\mathbf{b}) \leq F(\mathbf{b}),$$

we obtain

$$\lim_{n \to \infty} F_n(\mathbf{b}) < F(\mathbf{x}) + \epsilon.$$

We have selected $\mathbf{a}$ such that $a_j < x_j$ for all j; we now show that in addition $\mathbf{a}$ can be selected so that

$$\lim_{n \to \infty} F_n(\mathbf{a}) > F(\mathbf{x}) - \epsilon.$$

Indeed, since $\mathbf{x}$ is a continuity point of F there exists a $\mathbf{y} = (y_1, \cdots, y_k)$ such that

$$y_j < x_j, 1 \leq j \leq k \qquad \text{and} \qquad |\mathbf{x} - \mathbf{y}| < \delta$$

and hence

$$F(\mathbf{y}) > F(\mathbf{x}) - \epsilon.$$

Since D is dense, there is an $\mathbf{a} \in D$ such that

$$y_j < a_j < x_j, \qquad 1 \leq j \leq k,$$

and hence

$$\lim_{n \to \infty} F_n(\mathbf{a}) \geq F(\mathbf{y}) > F(\mathbf{x}) - \epsilon,$$

which verifies the inequality asserted above. Thus

$$\limsup_{n \to \infty} (F_n(\mathbf{x}) - F(\mathbf{x})) \leq \lim_{n \to \infty} F_n(\mathbf{b}) - F(\mathbf{x}) < \epsilon$$

and

$$\limsup_{n \to \infty} (F(\mathbf{x}) - F_n(\mathbf{x})) \leq F(\mathbf{x}) - \lim_{n \to \infty} F_n(\mathbf{a}) < \epsilon,$$

which proves that $F_n(\mathbf{x}) \to F(\mathbf{x})$ at all $\mathbf{x} \in \textit{Cont } F$. We next must prove that for every cell $(\mathbf{a}, \mathbf{b}] \subset E^{(k)}$, $\mu_F(\mathbf{a}, \mathbf{b}] \geq 0$. We first notice that since, for fixed $\mathbf{u} \in E^{(k)}$, $F(\mathbf{u} + t\mathbf{e})$ is nondecreasing in t (where $\mathbf{e} = (1, 1, \cdots, 1) \in E^{(k)}$), it is a *continuous function in* t at all but a countable set of values of t. We now show that F, *as a function over* $E^{(k)}$, is continuous at all points $\{\mathbf{u} + t\mathbf{e}, t \geq 0\}$ except for a countable number of them. Let $t_0 > 0$ be such that $\varphi(t) = F(\mathbf{u} + t\mathbf{e})$ is continuous at t_0. Suppose to the contrary that F is not continuous at $\mathbf{u} + t_0\mathbf{e}$. Then there is an $\epsilon > 0$ such that for every $\delta > 0$ there is a $\mathbf{y}$ such that

$$|\mathbf{y} - (\mathbf{u} + t_0\mathbf{e})| < \delta \qquad \text{and} \qquad |F(\mathbf{y}) - F(\mathbf{u} + t_0\mathbf{e})| \geq \epsilon.$$

Now there does exist a $\delta_0 > 0$ such that if $|t - t_0| < 2\delta_0$, then

$$|\varphi(t) - \varphi(t_0)| < \tfrac{1}{2}\epsilon.$$

Let us take $0 < \delta < \sqrt{\delta_0/k}$. Thus we have a cube

$$(\mathbf{u} + (t_0 - \delta_0)\mathbf{e}, \mathbf{u} + (t_0 + \delta_0)\mathbf{e}],$$

such that at the vertices, $\mathbf{u} + (t_0 - \delta_0)\mathbf{e}$ and $\mathbf{u} + (t_0 + \delta_0)\mathbf{e}$, φ differs from $\varphi(t_0)$ by an amount less than $\epsilon/2$, while there are points inside the sphere $\{\mathbf{z} \,\|\, \mathbf{z} - (\mathbf{u} + t_0\mathbf{e}) \,|\, < \delta\}$ which differ from $\varphi(t_0)$ by an amount greater than ϵ. Thus the monotonicity of F is violated, and we have proved that F, *as a function over $E^{(k)}$*, is continuous at all but a countable number of points of $\{\mathbf{u} + t\mathbf{e}, t \geq 0\}$. Hence there exists a sequence $\{t_m\}$, $t_m > 0$, and $t_m \to 0$ as $m \to \infty$ such that if

$$\mathbf{a}^{(m)} = \mathbf{a} + t_m\mathbf{e}, \ \mathbf{b}^{(m)} = \mathbf{b} + t_m\mathbf{e},$$

then *all vertices* of $(\mathbf{a}^{(m)}, \mathbf{b}^{(m)}]$ are points of continuity of F for *all m*. By what was proved above, namely, $F_n(\mathbf{x}) \to F(\mathbf{x})$ as $n \to \infty$ at all $\mathbf{x} \in Cont\ F$, it follows that $F_n \to F$ as $n \to \infty$ pointwise on the vertices of all cells $(\mathbf{a}^{(m)}, \mathbf{b}^{(m)}]$. By the definition of μ_F in Section 2.2,

$$\mu_{F_n}(\mathbf{a}^{(m)}, \mathbf{b}^{(m)}] \to \mu_F(\mathbf{a}^{(m)}, \mathbf{b}^{(m)}]$$

for every fixed m. But since F_n is a distribution function we have, by Theorem 1 of Section 2.2,

$$\mu_{F_n}(\mathbf{a}^{(m)}, \mathbf{b}^{(m)}] \geq 0.$$

Hence

$$\mu_F,(\mathbf{a}^{(m)}, \mathbf{b}^{(m)}] \geq 0$$

for all m. Since we have already shown that F is nondecreasing in each argument and is continuous from above, it follows that $\mu_F(\mathbf{a}, \mathbf{b}] \geq 0$. Thus we have shown that $F_n \xrightarrow{w} F$, where F is defined above. Finally we prove uniqueness. Let G be any other limit function such that $F_n \xrightarrow{w} G$, and let $\mathbf{x} \in E^{(k)}$. As was shown above, F and G are each continuous at all but a countable number of points in $\{\mathbf{x} + t\mathbf{e}, t > 0\}$. Hence (and here we invoke Lemma 1 for G) there exists a sequence $\{t_m\}$ such that $t_m > 0$, $t_m \to 0$ as $m \to \infty$, and such that F and G are continuous at each point in $\{\mathbf{x} + t_m\mathbf{e}, m = 1, 2, \cdots\}$. Hence, by what was proved above, for each m,

$$\lim_{n \to \infty} F_n(\mathbf{x} + t_m\mathbf{e}) = F(\mathbf{x} + t_m\mathbf{e}) = G(\mathbf{x} + t_m\mathbf{e}).$$

Since both F and G are continuous from above, it follows that $F(\mathbf{x}) = G(\mathbf{x})$. Q.E.D.

A stronger notion of convergence of distribution functions is the following.

Definition: If $\{F_n\}$ is a sequence of distribution functions over $E^{(k)}$, if

$$F_n \xrightarrow{w} F,$$

and if F is a distribution function over $E^{(k)}$, then we say that the sequence $\{F_n\}$ converges completely to F and write $F_n \xrightarrow{c} F$.

This type of convergence is what the entire subject of limiting distributions is concerned with. Its first important property appears in the corollary to Theorem 3.

Theorem 2 (Helly Compactness Theorem). Let $\{F_\lambda, \lambda \in \Lambda\}$ be an infinite set of distribution functions over $E^{(k)}$. Then there exists a sequence $\{\lambda_n\}$ of distinct elements in Λ such that $F_{\lambda_n} \xrightarrow{w}$.

Proof: Let D be a countable dense set in $E^{(k)}$, its points being designated by $\{\mathbf{x}^{(1)}, \mathbf{x}^{(2)}, \cdots\}$. By Theorem 1 we need only show that there exists a sequence $\{\lambda_n\}$ of distinct elements in Λ such that $\{F_{\lambda_n}\}$ converges pointwise over D. Since $\{F_\lambda(\mathbf{x}^{(1)}), \lambda \in \Lambda\} \subset [0, 1]$, then by the Bolzano-Weierstrass theorem there exists a sequence $\{\lambda(1, n)\}$ of distinct elements in Λ such that $\{F_{\lambda(1,n)}(\mathbf{x}^{(1)})\}$ converges; denote its limit by $F(\mathbf{x}^{(1)})$. Since

$$F_{\lambda(1,n)}(\mathbf{x}^{(2)}) \in [0, 1]$$

for all n, there exists a subsequence $\{\lambda(2, n)\} \subset \{\lambda(1, n)\}$ such that $\{F_{\lambda(2,n)}(\mathbf{x}^{(2)})\}$ converges; denote its limit by $F(\mathbf{x}^{(2)})$. Note that also

$$F_{\lambda(2,n)}(\mathbf{x}^{(1)}) \to F(\mathbf{x}^{(1)}) \qquad \text{as} \quad n \to \infty.$$

In general, for $j = 2, 3, \cdots$, we can select a subsequence $\{\lambda(j, n), n = 1, 2, \cdots\}$ of $\{\lambda(j - 1, n), n = 1, 2, \cdots\}$ such that

$$F_{\lambda(j,n)}(\mathbf{x}^{(m)}) \to F(\mathbf{x}^{(m)}) \qquad \text{as} \quad n \to \infty \quad \text{for} \quad m = 1, 2, \cdots, j.$$

Now select $\lambda_n = \lambda(n, n)$. Since

$$\{\lambda_m, \lambda_{m+1}, \cdots\} \subset \{\lambda(m, n), \qquad n = 1, 2, \cdots\}$$

for every m, we have

$$F_{\lambda_n}(\mathbf{x}^{(m)}) \to F(\mathbf{x}^{(m)}) \qquad \text{as} \quad n \to \infty \quad \text{for all } m,$$

which proves the theorem.

Theorem 3 (Helly–Bray Theorem). If $\{F_n\}$ is a sequence of distribution functions over $E^{(k)}$, if $F_n \xrightarrow{w} F$, if $I = [\mathbf{a}, \mathbf{b}]$ is a bounded, closed cell in $E^{(k)}$ whose boundaries contain no discontinuities of F, and if f is a continuous real- or complex-valued function over I, then

$$\int_I f(\mathbf{x})\, dF_n(\mathbf{x}) \to \int_I f(\mathbf{x})\, dF(\mathbf{x}) \qquad \text{as } n \to \infty.$$

Proof: Since $[\mathbf{a}, \mathbf{b}]$ is a closed, bounded set in $E^{(k)}$, then f is uniformly continuous over $[\mathbf{a}, \mathbf{b}]$. Thus, given $\epsilon > 0$, there is a $\delta_\epsilon > 0$ such that for x', $x'' \in [\mathbf{a}, \mathbf{b}]$ and $|\mathbf{x}' - \mathbf{x}''| < \delta_\epsilon$, then $|f(\mathbf{x}') - f(\mathbf{x}'')| < \frac{1}{3}\epsilon$. By Theorem 3 of Section 2.2, there exists a finite disjoint decomposition $\{I_j\}$ of I into cells such that the diameter of each cell is less than δ and no discontinuity of F is on any boundary point of I_j. Let $\mathbf{y}_j \in I_j$ and denote $a_j = f(\mathbf{y}_j)$. Then by the definition of μ_F in Section 2.2, by the definition of convergence in the wide sense, and by Lemma 2,

$$\mu_{F_n}(I_j) \to \mu_F(I_j)$$

as $n \to \infty$ for all j. Let $f_s = \sum a_j \chi_{I_j}$, where $\chi_{I_j}(\mathbf{x}) = 1$ if $\mathbf{x} \in I_j$ and $= 0$ if $\mathbf{x} \notin I_j$. Then

$$\left| \int_I f(\mathbf{x})\, dF_n(\mathbf{x}) - \int_I f(\mathbf{x})\, dF(\mathbf{x}) \right|$$

$$\leq \left| \int_I f(\mathbf{x})\, dF_n(\mathbf{x}) - \int_I f_s(\mathbf{x})\, dF_n(\mathbf{x}) \right|$$

$$+ \left| \int_I f_s(\mathbf{x})\, dF_n(\mathbf{x}) - \int_I f_s(\mathbf{x})\, dF(\mathbf{x}) \right|$$

$$+ \left| \int_I f_s(\mathbf{x})\, dF(\mathbf{x}) - \int_I f(\mathbf{x})\, dF(\mathbf{x}) \right|$$

$$< \tfrac{1}{3}\epsilon + \sum_j |a_j|\, |\mu_{F_n}(I_j) - \mu_F(I_j)| + \tfrac{1}{3}\epsilon.$$

For sufficiently large n, we apply Lemma 2 and obtain that the sum above is less than $\epsilon/3$, and thus the theorem is proved.

Corollary to Theorem 3 (also known as the Helly–Bray Theorem). If $\{F_n\}$ is a sequence of distribution functions over $E^{(k)}$, if $F_n \xrightarrow{c} F$, and if f is a bounded and continuous function over $E^{(k)}$, then

$$\int f(\mathbf{x}) \, dF_n(\mathbf{x}) \to \int f(\mathbf{x}) \, dF(\mathbf{x}) \qquad \text{as} \quad n \to \infty.$$

Proof: Let $\epsilon > 0$. Let $[\mathbf{a}, \mathbf{b}] = I$ be a cell in $E^{(k)}$ such that no discontinuity of F is on the boundary of $[\mathbf{a}, \mathbf{b}]$ and such that

$$\mu_F([\mathbf{a}, \mathbf{b}]) > 1 - \epsilon.$$

Let $K > 0$ be such that $|f(\mathbf{x})| \leq K$ for all x. Then by the Helly-Bray theorem,

$$\limsup_{n \to \infty} \left| \int f(\mathbf{x}) \, dF_n(\mathbf{x}) - \int f(\mathbf{x}) \, dF(\mathbf{x}) \right|$$

$$\leq \limsup_{n \to \infty} \left| \int_I f(\mathbf{x}) \, dF_n(\mathbf{x}) - \int_I f(\mathbf{x}) \, dF(\mathbf{x}) \right|$$

$$+ \limsup_{n \to \infty} \left| \int_{E^{(k)} \setminus I} f(\mathbf{x}) \, dF_n(\mathbf{x}) - \int_{E^{(k)} \setminus I} f(\mathbf{x}) \, dF(\mathbf{x}) \right|$$

$$\leq 2K\epsilon,$$

from which we obtain the conclusion of the corollary.

The following theorem is not only of interest in its own right but will be applied in Chapter 6. For simplicity, we deal only over $E^{(1)}$; the same arguments hold over $E^{(k)}$. If $\{H_n\}$ is a bounded sequence of nondecreasing functions over $(-\infty, +\infty)$, then in the same way in which the Helly compactness theorem was proved one can prove that there exists a subsequence $\{H_{n_k}\}$ of $\{H_n\}$ and a bounded, nondecreasing function H such that $H_{n_k}(x) \to H(x)$ as $k \to \infty$ for all $x \in Cont\ H$. In this case we also write $H_{n_k} \xrightarrow{w} H$, and we say that $\{H_{n_k}\}$ converges to H *in the wide sense.* If, in addition,

$$\int dH_{n_k}(x) \to \int dH(x) \qquad \text{as} \quad k \to \infty$$

(that is, if total variations converge to total variation), then we write $H_{n_k} \xrightarrow{c} H$ and say that $\{H_{n_k}\}$ *converges completely to H.*

Theorem 4. Let $\{H_n\}$ be a bounded sequence of nondecreasing functions. A necessary and sufficient condition that every subsequence of

$\{H_n\}$ which converges in the wide sense also converges completely is that

$$\int_{|x| \geq T} dH_n(x) \to 0 \qquad \text{as} \quad T \to \infty$$

uniformly in n.

Proof: In order to show that the condition is sufficient, let $\{H_{n_k}\}$ be a subsequence of $\{H_n\}$ such that $H_{n_k} \xrightarrow{w} H$; we shall prove $H_{n_k} \xrightarrow{c} H$ as $k \to \infty$. Let $\epsilon > 0$. Let $\pm x_0 \in Cont\, H$ be such that

$$\int_{|x| \geq x_0} dH_{n_k}(x) < \tfrac{1}{3}\epsilon \qquad \text{for all } k,$$

and such that

$$\int_{|x| \geq x_0} dH(x) < \tfrac{1}{3}\epsilon.$$

Then for all sufficiently large k,

$$\left| \int dH_{n_k}(x) - \int dH(x) \right|$$

$$\leq \left| \int_{|x| \geq x_0} dH_{n_k}(x) \right| + \left| \int_{|x| < x_0} dH_{n_k}(x) - \int_{|x| < x_0} dH(x) \right|$$

$$+ \left| \int_{|x| \geq x_0} dH(x) \right| < \epsilon,$$

which concludes the proof that the condition is sufficient. Conversely, suppose that every subsequence which converges in the wide sense converges completely; we wish to prove that

$$\int_{|x| \geq T} dH_n(x) \to 0 \qquad \text{as} \quad T \to \infty$$

uniformly in n, that is, to prove that for every $\epsilon > 0$ there exists a $T_\epsilon > 0$ such that if $T > T_\epsilon$, then

$$\int_{|x| \geq T} dH_n(x) < \epsilon \qquad \text{for all } n.$$

Let us suppose that this conclusion were not so. Then there is an $\epsilon > 0$

such that for every $t > 0$ there is a $T > t$ and an integer n such that

$$\int_{|x| \geq T} dH_n(x) \geq 2\epsilon.$$

Thus there exists an increasing sequence $\{T_k\}$ and an increasing sequence of integers $\{n_k\}$ such that $T_k \to \infty$ as $k \to \infty$ and such that

$$\int_{|x| \geq T_k} dH_{n_k}(x) \geq 2\epsilon.$$

By the Helly compactness theorem there is a subsequence $\{H_{m_k}\}$ of $\{H_{n_k}\}$ such that $H_{m_k} \xrightarrow{w} H$, where H is some bounded nondecreasing function. We shall arrive at a contradiction by showing that it is not true that $H_{m_k} \xrightarrow{c} H$; that is, we shall show that $\{ \int dH_{m_k}(x) \}$ does not converge to $\int dH(x)$. Indeed, let $K > 0$, $\pm K \in Cont\ H$ and $\pm K \in Cont\ H_{m_k}$ for all k. Let K be so large that

$$\int_{|x| \geq K} dH(x) < \tfrac{1}{2}\epsilon.$$

Since $H_{m_k} \xrightarrow{w} H$, then for all sufficiently large k, we have

$$\left| \int_{|x| < K} dH_{m_k}(x) - \int_{|x| < K} dH(x) \right| < \tfrac{1}{2}\epsilon.$$

Hence, for all sufficiently large k,

$$\int dH_{m_k}(x) - \int dH(x)$$

$$= \int_{|x| \geq K} dH_{m_k}(x) + \left(\int_{|x| < K} dH_{m_k}(x) - \int_{|x| < K} dH(x) \right)$$

$$- \int_{|x| \geq K} dH(x) \geq 2\epsilon - \tfrac{1}{2}\epsilon - \tfrac{1}{2}\epsilon = \epsilon,$$

which concludes the proof.

EXERCISES

1. Prove Theorem 1 quickly when $k = 1$.
2. Prove: If $\{F_n\}$ is a sequence of distribution functions over $E^{(k)}$, if $F_n \xrightarrow{c} F$, and if F is continuous, then $F_n \to F$ uniformly.

3. Let $\{F_n\}$ be a sequence of distribution functions over $E^{(1)}$, and let $\{Q_n\}$ be the sequence of corresponding concentration functions. For every n and every $L \geq 0$, let $x_L^{(n)}$ satisfy

$$Q_n(L) = F_n\,(x_L^{(n)} + L) - F_n(x_L^{(n)} - 0).$$

(See Exercise 4 in Section 3.2.) Prove that if every subsequence of $\{F_n\}$ which converges in the wide sense converges completely, then, for every fixed $L > 0$, the sequence $\{x_L^{(n)}\}$ is bounded.

4.2. The Continuity Theorem

Almost all limit theorems for distribution functions are proved by use of the continuity theorem. We have already proved the uniqueness theorem which states that there is a one-to-one correspondence between the set of all characteristic functions over $E^{(k)}$ and the set of all distribution functions over $E^{(k)}$. The continuity theorem essentially states that this one-to-one mapping is bicontinuous. In this section we prove the continuity theorem, and an application of it is made to prove a very useful theorem.

Theorem 1. If F, F_1, F_2, $\cdots$ are distribution functions over $E^{(k)}$ with corresponding characteristic functions f, f_1, f_2, $\cdots$, and if $F_n \xrightarrow{c} F$, then $f_n(\mathbf{u}) \to f(\mathbf{u})$ at all $\mathbf{u} \in E^{(k)}$.

Proof: This is an easy application of the corollary to the Helly-Bray theorem (Theorem 3 in Section 4.1). For each fixed $\mathbf{u} \in E^{(k)}$, $\exp[i(\mathbf{u}, \mathbf{x})]$ is a bounded, continuous function of $\mathbf{x}$. Hence

$$\int \exp[i(\mathbf{u}, \mathbf{x})]\,dF_n(\mathbf{x}) \to \int \exp[i(\mathbf{u}, \mathbf{x})]\,dF(\mathbf{x}),$$

or $f_n(\mathbf{u}) \to f(\mathbf{u})$ as $n \to \infty$ at every $\mathbf{u} \in E^{(k)}$.

Theorem 1 is used on occasion; however, the most frequently used half of the continuity theorem is Theorem 2.

Theorem 2 (The Continuity Theorem). If $\{f_n\}$ is a sequence of characteristic functions of the corresponding distribution functions $\{F_n\}$ over $E^{(k)}$, and if $\{f_n\}$ converges pointwise to some function f which is continuous at $(0, \cdots, 0) \in E^{(k)}$, then f is the characteristic function of a distribution function F and $F_n \overset{c}{\to} F$.

Proof: By the Helly compactness theorem there exists a subsequence $\{F_{n_j}\}$ of $\{F_n\}$ such that $F_{n_j} \overset{w}{\to}$, say $F_{n_j} \overset{w}{\to} F$ as $j \to \infty$. We shall prove that F is a distribution function. We assume to the contrary that

$$\mu_F(E^{(k)}) = \Delta < 1.$$

Let $0 < \epsilon < 1 - \Delta$. Since $f(0) = 1$, and since f is continuous at the origin, then for $\tau > 0$ sufficiently small,

$$\left| (2\tau)^{-k} \int_{-\tau}^{\tau} \cdots \int_{-\tau}^{\tau} (f(t) - 1) \, dt \right| < \tfrac{1}{2}\epsilon,$$

or, what amounts to the same

$$\left| (2\tau)^{-k} \int_{-\tau}^{\tau} \cdots \int_{-\tau}^{\tau} f(t) \, dt \right| > 1 - \tfrac{1}{2}\epsilon > \Delta + \tfrac{1}{2}\epsilon.$$

Let $M > 0$ and denote $\mathbf{M} = (M, \cdots, M) \in E^{(k)}$. Let $I = [-\mathbf{M}, \mathbf{M}]$, and let M be large enough so that $1/\tau M < \epsilon/4$ and so that no hyperplane of discontinuity of any F_n or F contains a face of I. Now by the Fubini theorem,

$$\int_{-\tau}^{\tau} \cdots \int_{-\tau}^{\tau} f_{n_j}(t) \, dt = \int \left[\int_{-\tau}^{\tau} \cdots \int_{-\tau}^{\tau} \exp \left[i(t, x) \right] dt \right] dF_{n_j}(x)$$

$$= \int \left(\prod_{m=1}^{k} \frac{2}{x_m} \sin \tau x_m \right) dF_{n_j}(x).$$

Let us denote the integrand of this last integral by $J(x)$. If $x \notin I$, then

$$\max \{ |x_m| \, | \, 1 \le m \le k \} > M,$$

and thus

$$|J(x)| \le (2\tau)^{k-1}(2/M).$$

For $x \in I$,

$$|J(x)| \le (2\tau)^k.$$

Let us denote

$$K_j = \left(\frac{1}{2\tau} \right)^k \int_{-\tau}^{\tau} \cdots \int_{-\tau}^{\tau} f_{n_j}(t) \, dt.$$

Then for sufficiently large values of j,

$$| K_j | = \left(\frac{1}{2\tau}\right)^k \left| \int J(\mathbf{x}) \, dF_{n_j}(\mathbf{x}) \right|$$

$$\leq \left(\frac{1}{2\tau}\right)^k \left(\int_I | J(\mathbf{x}) | \, dF_{n_j}(\mathbf{x}) + \int_{E^k \setminus I} | J(\mathbf{x}) | \, dF_{n_j}(\mathbf{x}) \right)$$

$$\leq \mu_{F_{n_j}}(I) + 1/\tau M < \Delta + \tfrac{1}{2}\epsilon.$$

On the other hand, by the Lebesgue dominated convergence theorem and the inequality first established in this proof,

$$| K_j | \to | (2\tau)^{-k} \int_{-\tau}^{\tau} \cdots \int_{-\tau}^{\tau} f(\mathbf{t}) \, d\mathbf{t} | > \Delta + \tfrac{1}{2}\epsilon,$$

which yields a contradiction. Hence $\Delta = 1$, and $F_{n_j} \xrightarrow{c} F$. We now show that $F_n \xrightarrow{c} F$. We do this by showing that every subsequence which converges in the wide sense also converges completely to F (since a bounded sequence of real numbers converges to a number L if and only if every convergent subsequence converges to L). Accordingly, let $\{F_{m_i}\}$ be any other subsequence of $\{F_n\}$ which converges in the wide sense. Then, by what was proved above, there exists a distribution function H such that $F_{m_i} \xrightarrow{c} H$. Let φ and ψ denote the characteristic functions of F and H respectively. By Theorem 1, $f_{n_j} \to \varphi$ and $f_{m_i} \to \psi$ pointwise as $j \to \infty$. But $\{ f_{n_j} \}$ and $\{ f_{m_i} \}$ are both subsequences of $\{ f_n \}$, which by hypothesis converges pointwise to f. Hence $f = \varphi = \psi$, and by the uniqueness theorem, $F = H$. Q.E.D.

It should be noticed that the assumption of continuity is needed in the hypothesis of Theorem 2. Indeed, consider

$$f_n(u) = \exp(-nu^2/2)$$

for all $u \in E^{(1)}$. Then f_n is the characteristic function of a $\mathfrak{N}(0, n)$ distribution function. Now $\lim_{n \to \infty} f_n(u) = 0$ if $u \neq 0$ and $= 1$ if $u = 0$. Thus it can occur that the limit of a sequence of characteristic functions is not continuous, and therefore not a characteristic function.

We now record an easy consequence of the continuity theorem which has many applications.

Theorem 3. Let $T : E^{(k)} \to E^{(m)}$ be a continuous mapping, and let $\mathbf{X}, \mathbf{X}^{(1)}, \mathbf{X}^{(2)}, \cdots$ be a sequence of k-dimensional random variables such that

$$F_{X^{(n)}} \xrightarrow{c} F_X.$$

Then

$$F_{TX^{(n)}} \xrightarrow{c} F_{TX}.$$

Proof: Let

$$\mathbf{v} = (v_1, \cdots, v_m) \in E^{(m)}.$$

By Theorem 2 we need only prove that

$$E(\exp i(\mathbf{v}, T\mathbf{X}^{(n)})) \to E(\exp i(\mathbf{v}, T\mathbf{X}))$$

as $n \to \infty$, for all $\mathbf{v} \in E^{(m)}$. But by Theorem 1 of Section 2.4 and the corollary to the Helly–Bray theorem,

$$E(\exp i(\mathbf{v}, T\mathbf{X}^{(n)}))$$

$$= \int \cdots \int \exp \left[i(\mathbf{v}, T(\mathbf{z})) \right] dF_{\mathbf{X}^{(n)}}(\mathbf{z})$$

$$\to \int \cdots \int \exp \left[i(\mathbf{v}, T(\mathbf{z})) \right] dF_{\mathbf{X}}(\mathbf{z}) \qquad \text{as } n \to \infty$$

$$= E(\exp i(\mathbf{v}, T\mathbf{X})). \quad \text{Q.E.D.}$$

This theorem considerably generalizes theorems of the following type: if $X, X_1, X_2, \cdots$ and $Y, Y_1, Y_2, \cdots$ are two sequences of random variables such that $F_{X_n} \xrightarrow{c} F_X$ and $F_{Y_n} \xrightarrow{c} F_Y$, and if $P[Y = K] = 1$ for some constant K, then $F_{X_n+Y_n} \xrightarrow{c} F_{X+Y}$. Indeed, it is easy to verify in this case (see Exercise 1) that the joint distribution function of (X_n, Y_n) converges completely to that of (X, Y). In this case $k = 2$, $m = 1$ and $T(x, y) = x + y$, which is continuous, and the conclusion follows from Theorem 3.

EXERCISES

1: Prove: If $X, X_1, X_2, \cdots$ and $Y, Y_1, Y_2, \cdots$ are two sequences of random variables such that $F_{X_n} \xrightarrow{c} F_X$, $F_{Y_n} \xrightarrow{c} F_Y$, and $P[Y = 1] = 1$, then

$$F_{X_n, Y_n} \xrightarrow{c} F_{X, Y}.$$

2: Let $\{U_n\}$ be a sequence of random variables such that the distribution of U_n is $\mathcal{P}(n)$. Let

$$X_n = (U_n - n)/\sqrt{n}.$$

Prove that $F_{X_n} \xrightarrow{c} F$, where F is the $\mathfrak{N}(0, 1)$ distribution function. (Hint: use the continuity theorem.)

3: Let X, X_1, X_2, $\cdots$ be a sequence of random variables with F, F_1, F_2, $\cdots$ as the corresponding sequence of distribution functions. If

$$\mu_{F_n}(A) \to \mu_F(A)$$

over all Borel sets A in $E^{(1)}$, and if φ is any Borel-measurable function over $(-\infty, +\infty)$, then

$$F_{\varphi(X_n)}(x) \to F_{\varphi(X)}(x)$$

for *all* $x \in E^{(1)}$.

4: Let X, X_1, X_2, $\cdots$ and Y, Y_1, Y_2, $\cdots$ be two sequences of random variables such that X and Y are independent, X_n and Y_n are independent for every n, $F_{X_n} \xrightarrow{c} F_X$, and $F_{Y_n} \xrightarrow{c} F_Y$. Give two proofs that $F_{X_n+Y_n} \xrightarrow{c} F_{X+Y}$.

4.3. Refinements of the Continuity Theorem for Nonvanishing Characteristic Functions

We shall eventually wish to take the logarithm of a characteristic function that never vanishes. Also, in case $\{f_n\}$ is a sequence of nonvanishing characteristic functions that converges pointwise to a nonvanishing characteristic function, we shall eventually wish to state that this implies that $\log f_n \to \log f$ pointwise as $n \to \infty$. This section is devoted to the attainment of an unambiguous definition of such logarithms and the proof of the limit theorem just described.

We shall deal only with distributions over $E^{(1)}$. The problem here arises from the fact that if $f(u)$ is a nonvanishing characteristic function, then

$$\log f(u) = \log |f(u)| + i\varphi(u),$$

where $\varphi(u)$ is real but only uniquely determined modulo 2π for each u. Now, we shall want $\log f(u)$ to be continuous, and hence $\varphi(u)$ will have to be continuous. If, say, $f(u) = e^{iu}$, we shall want to write $\log f(u) = iu$ and not $\log f(u) = i\{u(\mathrm{mod}\ 2\pi)\}$. Thus we hope that there exists a continuous real-valued function $\varphi(u)$ such that $\varphi(0) = 0$ and such that

$$\log f(u) = \log |f(u)| + i\varphi(u).$$

In addition, we hope that there is just one such $\varphi(u)$. Once this is done, then we need to prove that if $f, f_1, f_2, \cdots$ is a sequence of nonvanishing characteristic functions, and if $f_n(u) \to f(u)$ at all u, then

$$\log f_n(u) \to \log f(u) \qquad \text{as} \quad n \to \infty$$

at all u. If

$$\log f_n(u) = \log |f_n(u)| + i\varphi_n(u)$$

and

$$\log f(u) = \log |f(u)| + i\varphi(u),$$

then the problem of proving $\log f_n(u) \to \log f(u)$ as $n \to \infty$ for all u becomes that of proving $\varphi_n(u) \to \varphi(u)$ as $n \to \infty$ at all real u.

Theorem 1. Let h be a continuous complex-valued function defined over $(-\infty, +\infty)$ such that $h(t) \neq 0$ for all $t \in (-\infty, +\infty)$ and $h(0) = 1$. Then there exists one and only one continuous real-valued function $\varphi(t)$ such that $\varphi(0) = 0$ and

$$h(t) = |h(t)| \exp i\varphi(t)$$

for all t.

Proof: We first prove the uniqueness of φ. Suppose φ_1 and φ_2 are two such functions. Then there is an integer-valued function $k(t)$ such that

$$\varphi_1(t) = \varphi_2(t) + 2\pi k(t).$$

Since φ_1 and φ_2 are continuous, then $k(t)$ is continuous. But by hypothesis,

$$\varphi_1(0) = \varphi_2(0) = 0;$$

hence $k(0) = 0$, which implies $k(t) = 0$ for all t. We now show that there exists such a $\varphi(t)$. By the uniqueness of φ it suffices to prove the existence of φ on every bounded interval $[0, M]$. Since h is continuous,

$$\min \{ |h(u)| \mid u \in [0, M] \} = c > 0.$$

Without loss of generality we may assume that $|h(u)| \equiv 1$ over $[0, M]$. Now h being continuous over $[0, M]$ implies that it is uniformly continuous; hence there is a $\delta > 0$ such that if $0 \leq t' < t'' \leq M$ and $0 < t'' -$

$t' < \delta$, then $| h(t') - h(t'') | < 1$. Clearly (because $| h(t) | \equiv 1$), if $| t'' - t' | < \delta$, then

$$| \text{Arg} (h(t')/h(t'')) | < \tfrac{1}{2}\pi,$$

where $\text{Arg}\, z$ means $\text{Arctan}\, \{\text{Im}\, z/\text{Re}\, z\}$. We select

$$0 = t_0 < t_1 < \cdots < t_m = M$$

such that

$$0 < \max \{t_j - t_{j-1} \,|\, 1 \leq j \leq m\} < \delta.$$

For $t \in [t_0, t_1]$, we define

$$\varphi(t) = \text{Arg}\, h(t).$$

Clearly $\varphi(t)$ is continuous in $[t_0, t_1]$, since

$$\varphi(t) = \text{Arctan}\, \{\text{Im}\, h(t)/\text{Re}\, h(t)\},$$

both the numerator and denominator of the ratio are continuous, and $h(t) \neq 0$ over $[t_0, t_1]$. If $\varphi(t)$ has been defined over $[t_0, t_j]$, we define φ in $[t_j, t_{j+1}]$ by

$$\varphi(t) = \varphi(t_j) + \text{Arg}\, (h(t)/h(t_j)),$$

where again the argument is well-defined. Q.E.D.

We recall the following definition.

Definition: Let $\mathfrak{F} = \{ f_\lambda, \lambda \in \Lambda \}$ be a family of functions defined over some metric space $\mathfrak{X}$ and whose values are in some metric space $\mathfrak{Y}$. Then $\mathfrak{F}$ is said to be uniformly equicontinuous if for every $\epsilon > 0$ there is a $\delta_\epsilon > 0$ such that if $x', x'' \in \mathfrak{X}$ and $\text{dist}\, (x', x'') < \delta_\epsilon$, then

$$\text{dist}\, (f_\lambda(x'), f_\lambda(x'')) < \epsilon$$

for all $\lambda \in \Lambda$.

Theorem 2. If $\{ f_n \}$ is a sequence of characteristic functions and if $f_n(u) \to f(u)$ as $n \to \infty$ at all u, where f is a characteristic function, then $\{ f, f_1, f_2, \cdots \}$ are uniformly equicontinuous.

Proof: Let $\epsilon > 0$; we shall show that there exists a $\delta_\epsilon > 0$ such that $| h | < \delta_\epsilon$ implies that

$$| f_n(u + h) - f_n(u) | < \epsilon$$

for all u and all n. By the continuity theorem, $F_n \xrightarrow{c} F$, where F_n and F are the distribution functions determined by f_n and f respectively. By Theorem 4 in Section 4.1 there exists a $K > 0$ such that, for all n,

$$\int_{|x|\geq K} dF_n(x) < \tfrac{1}{4}\epsilon.$$

Let $\delta_\epsilon > 0$ satisfy the inequalities

$$0 < \delta_\epsilon K < \tfrac{1}{2}\pi \qquad \text{and} \qquad |\exp(i\delta_\epsilon K) - 1| < \tfrac{1}{2}\epsilon.$$

Then it easily follows that if $0 \leq |h| < \delta_\epsilon$ and $0 \leq |x| < K$, then $|\exp(ihx) - 1| < \epsilon/2$. Thus, for every u and n.

$$|f_n(u+h) - f_n(u)|$$

$$\leq \int |\exp[i(u+h)x] - \exp(iux)| \, dF_n(x)$$

$$= \int |\exp(ihx) - 1| \, dF_n(x)$$

$$\leq \int_{-K}^{K} |\exp(ihx) - 1| \, dF_n(x) + 2\int_{|x|\geq K} dF_n(x)$$

$$< \epsilon \qquad \text{for} \quad |h| < \delta_\epsilon.$$

Thus $\{f_1, f_2, \cdots\}$ are equicontinuous, and because of the uniform continuity of f, so are $\{f, f_1, f_2, \cdots\}$.

The next theorem strengthens the conclusion of the continuity theorem.

Theorem 3. If $\{f, f_1, f_2, \cdots\}$ are characteristic functions, and if $f_n(u) \to f(u)$ as $n \to \infty$ at all u, then $f_n \to f$ uniformly over every bounded interval.

Proof: Let $[a, b]$ be any bounded interval, and let $\epsilon > 0$ be arbitrary. By Theorem 2, $\{f, f_1, f_2, \cdots\}$ are uniformly equicontinuous, so there exists a $\delta_\epsilon > 0$ such that if $|u' - u''| < \delta_\epsilon$, then

$$|f(u') - f(u'')| < \tfrac{1}{3}\epsilon \qquad \text{and} \qquad |f_n(u') - f_n(u'')| < \tfrac{1}{3}\epsilon$$

for all n. Let

$$a = t_0 < t_1 < \cdots < t_m = b$$

be such that

$$0 < t_j - t_{j-1} < \delta_\epsilon$$

for $1 \leq j \leq m$. Then there exists an integer N_ϵ such that $n > N_\epsilon$ implies that

$$|f_n(t_j) - f(t_j)| < \tfrac{1}{3}\epsilon$$

for $0 \leq j \leq m$. Let $u \in [a, b]$, and let j be such that $t_{j-1} \leq u \leq t_j$. Then for $n > N_\epsilon$,

$$|f_n(u) - f(u)| \leq |f_n(u) - f_n(t_j)| + |f_n(t_j) - f(t_j)|$$
$$+ |f(t_j) - f(u)| < \epsilon,$$

which establishes the conclusion.

Theorem 1 above entitles us to establish the following definition.

Definition: If $h(u)$ is a nonvanishing, continuous, complex-valued function over $(-\infty, +\infty)$, and if $h(0) = 1$, then $\log h(u)$ is defined as $\log |h(u)| + i\varphi(u)$, where $\varphi(u)$ is that unique function determined by Theorem 1 which is continuous, real-valued, and satisfies $\varphi(0) = 0$ and $h(u) = |h(u)| \exp(i\varphi(u))$.

The principal theorem of this section which is used repeatedly in Chapter 6 is Theorem 4.

Theorem 4. If $\{f, f_1, f_2, \cdots\}$ are characteristic functions that never vanish, then $f_n(u) \to f(u)$ as $n \to \infty$ if and only if $\log f_n(u) \to \log f(u)$ as $n \to \infty$, at all $u \in (-\infty, +\infty)$.

Proof: The "if" part of the theorem easily follows from the definition of logarithm and by taking exponents. We therefore assume now that $f_n \to f$ pointwise and shall prove $\log f_n \to \log f$ pointwise. By Theorem 1 there exist unique real-valued continuous functions $\{\varphi, \varphi_1, \varphi_2, \cdots\}$ which vanish at 0 and such that

$$f(u) = |f(u)| \exp i\varphi(u) \qquad \text{and} \qquad f_n(u) = |f_n(u)| \exp i\varphi_n(u)$$

for all u and all n. We need only prove $\varphi_n(u) \to \varphi(u)$ as $n \to \infty$ over every bounded interval of which one endpoint is 0. Let $[0, T]$ be such an interval, and define ζ by

$$\zeta = \inf \{ \, |f_n(u)| \, , \, |f(u)| \, \| \, n = 1, 2, \cdots, u \in [0, T] \}.$$

We first prove that $\zeta > 0$. Since f and every f_n do not vanish and are continuous over $[0, T]$, it is sufficient to prove that, for some N,

$$\inf \{ \, |f_n(u)| \, \| \, n \geq N, \, u \in [0, T] \} > 0.$$

Accordingly, let

$$\xi = \min \{ \, |f(u)| \, \| \, u \in [0, T] \}.$$

Since $f(u) \neq 0$ for all $u \in [0, T]$, and since f is continuous, then $\xi > 0$. Since $f_n \to f$, then by Theorem 3, $f_n \to f$ uniformly over $[0, T]$; that is, there exists an N such that $n \geq N$ implies that

$$|f_n(u) - f(u)| < \tfrac{1}{2}\xi$$

for all $u \in [0, T]$. Hence, for all $n \geq N$,

$$|f_n(u)| = |f(u) - (f(u) - f_n(u))| \geq |f(u)| - |f(u) - f_n(u)|$$
$$> \xi - \tfrac{1}{2}\xi = \tfrac{1}{2}\xi \qquad \text{for all} \quad u \in [0, T].$$

Thus we have proved that $\zeta > 0$. Now let $0 < \epsilon < \tfrac{1}{2}\zeta$. Since, by Theorem 2, $\{ f, f_1, f_2, \cdots \}$ are uniformly equicontinuous, there is a $\delta_\epsilon > 0$ such that if $|u' - u''| < \delta_\epsilon$, then

$$|f(u') - f(u'')| < \epsilon$$

and, for all n,

$$|f_n(u') - f_n(u'')| < \epsilon.$$

Let

$$0 = t_0 < t_1 < \cdots < t_m = T$$

be such that

$$t_i - t_{i-1} < \delta_\epsilon$$

for $i = 0, 1, \cdots, m$. We first prove that $\varphi_n(u) \to \varphi(u)$ as $n \to \infty$ for all $u \in [t_0, t_1]$. In order to do this we first prove that

$$|\varphi(u)| < \tfrac{1}{3}\pi \qquad \text{and} \qquad |\varphi_n(u)| < \tfrac{1}{3}\pi$$

for all $u \in [t_0, t_1]$ and all n. We need only show that for all $u \in [t_0, t_1]$ and all n that

$$|\exp i\varphi_n(u) - 1| < 1 \qquad \text{and} \qquad |\exp i\varphi(u) - 1| < 1.$$

We first observe that $\{ |f|, |f_1|, |f_2|, \cdots \}$ are also uniformly equicontinuous, with the same δ_ϵ, because of the inequality $||\,a\,|\,-\,|\,b\,||\,\le\,|\,a\,-\,b\,|$. Then, for all $u \in [t_0,\, t_1]$, and since $f_n(0) = |\,f_n(0)\,| = 1$, we have

$$|\exp i\varphi_n(u)\, -\, 1\,| \le (\,|\,f_n(u)\,|/\zeta)\,|\exp i\varphi_n(u)\, -\, 1\,|$$

$$= (1/\zeta)\,|\,f_n(u)\, -\, |\,f_n(u)\,|\,|$$

$$\le (1/\zeta)\,(\,|\,f_n(u)\, -\, f_n(0)\,|\, +\, ||\,f_n(0)\,|\, -\, |\,f_n(u)\,|\,|)$$

$$< 2\epsilon/\zeta < 1.$$

Hence $|\,\varphi_n(u)\,| < \pi/3$ for all n, $|\,\varphi(u)\,| < \pi/3$, all inequalities holding for all $u \in [t_0,\, t_1]$. Since Arctan, "imaginary part of," and "real part of" are continuous functions, and since $\mathrm{Re}\,f(u) \ne 0$ and $\mathrm{Re}\,f_n(u) \ne 0$ for all $u \in [t_0,\, t_1]$, we have

$$\varphi_n(u) = \mathrm{Arctan}\,\frac{\mathrm{Im}\,f_n(u)}{\mathrm{Re}\,f_n(u)} \to \mathrm{Arctan}\,\frac{\mathrm{Im}\,f(u)}{\mathrm{Re}\,f(u)}$$

as $n \to \infty$ for all $u \in [t_0,\, t_1]$.

Now assume that we have proved that $\varphi_n(u) \to \varphi(u)$ as $n \to \infty$ for all $u \in [t_0,\, t_j]$; we shall show that this implies that $\varphi_n(u) \to \varphi(u)$ as $n \to \infty$ for all $u \in [t_j,\, t_{j+1}]$. In exactly the same manner as above, it can be proved that

$$|\,\varphi_n(u)\, -\, \varphi_n(t_j)\,| < \tfrac{1}{3}\pi, \qquad |\,\varphi(u)\, -\, \varphi(t_j)\,| < \tfrac{1}{3}\pi$$

for all $u \in [t_j,\, t_{j+1}]$. Now, for $u \in [t_j,\, t_{j+1}]$,

$$f_n(u)\,\exp\{-i\varphi_n(t_j)\} \to f(u)\,\exp\{-i\varphi(t_j)\}$$

as $n \to \infty$, from which we obtain, as before,

$$\varphi_n(u)\, -\, \varphi_n(t_j) \to \varphi(u)\, -\, \varphi(t_j)$$

as $n \to \infty$. Thus, the theorem is proved.

EXERCISES

1. Let $f(u)$, $g(u)$ be two nonvanishing, continuous complex-valued functions defined over $(-\infty,\, +\infty)$ such that $f(0) = g(0) = 1$. Prove that

$$\log\,(\,f(u)g(u)) = \log f(u)\, +\, \log g(u).$$

2. Prove that if $\{ f, f_1, f_2, \cdots \}$ are characteristic functions which never vanish, and if $f_n(u) \to f(u)$ as $n \to \infty$ at all $u \in (-\infty, +\infty)$, then $\log f_n \to \log f$ uniformly over every bounded interval.

3. Verify that if φ_1, φ_2 are continuous real-valued functions defined over $(-\infty, +\infty)$, if $\varphi_1(0) = \varphi_2(0) = 0$, and if $(\varphi_1(u) - \varphi_2(u))/2\pi$ is an integer for every real u, then $\varphi_1 = \varphi_2$.

4.4. The Four Types of Convergence: Almost Sure, in Law, in Probability, and in rth Mean

In this section the four basic types of convergence used in probability theory will be defined and investigated.

Definition: If $\mathbf{X}$ is a vector random variable, then by the **law** of $\mathbf{X}$, $\mathcal{L}(\mathbf{X})$, we mean its joint distribution function or joint characteristic function or (in cases of absolute continuity) its joint density function.

If, for example, we write "$\mathcal{L}(X)$ is $\mathfrak{N}(\mu, \sigma^2)$," we mean that the distribution function of X is normal with expectation equal to μ and variance equal to σ^2.

Definition: If $\mathbf{X}$, $\mathbf{X}^{(1)}$, $\mathbf{X}^{(2)}$, $\cdots$ are k-dimensional random variables, we say that $\{\mathbf{X}^{(n)}\}$ converges in law to $\mathbf{X}$ and write $\mathcal{L}(\mathbf{X}^{(n)}) \to \mathcal{L}(\mathbf{X})$ as $n \to \infty$ if $F_{\mathbf{X}}{}^{(n)} \xrightarrow{c} F_{\mathbf{X}}$.

Definition: If $\{\mathbf{X}^{(n)}\}$ is a sequence of k-dimensional vector random variables, where

$$\mathbf{X}^{(n)} = (X_1^{(n)}, \cdots, X_k^{(n)}),$$

we define $[\mathbf{X}^{(n)} \to] = \{\omega \in \Omega \mid \{X_j^{(n)}(\omega)\}$ is a convergent sequence of numbers, $1 \leq j \leq k\}$. We say $\{\mathbf{X}^{(n)}\}$ converges almost surely, and write $\mathbf{X}^{(n)} \xrightarrow{\text{a.s.}}$ or $\{\mathbf{X}^{(n)}\}$ converges almost surely, if

$$P[\mathbf{X}^{(n)} \to] = 1.$$

It should be noticed that *almost sure convergence* just defined is usually referred to as *almost everywhere* convergence in measure theory.

Proposition 1. If $\{\mathbf{X}^{(n)}\}$ is a sequence of k-dimensional random variables, and if $\mathbf{X}^{(n)} \overset{\text{a.s.}}{\to}$, then there exists a k-dimensional random variable $\mathbf{X}$, uniquely determined except over an event of probability zero, such that

$$P\{\omega \mid \mathbf{X}^{(n)}(\omega) \to \mathbf{X}(\omega)\} = 1 \qquad (\text{or } P[\mathbf{X}^{(n)} \to \mathbf{X}] = 1).$$

Proof: For every $j \in \{1, 2, \cdots, k\}$, define

$$X_j = \inf_{n \geq 1} \sup_{r \geq n} X_j^{(r)}.$$

Clearly, X_j is a random variable uniquely defined over an event of probability one. Letting $\mathbf{X} = (X_1, \cdots, X_k)$, we obtain the conclusion.

If $\{\mathbf{X}^{(n)}\}$ is a sequence of k-dimensional vector random variables which converges almost surely, and if $\mathbf{X}$ is the k-dimensional random vector whose existence is guaranteed by the above proposition, then we write

$$\mathbf{X}^{(n)} \overset{\text{a.s.}}{\to} \mathbf{X} \qquad \text{or} \qquad \mathbf{X}^{(n)} \to \mathbf{X} \text{ a.s.}$$

In what follows, if $\mathbf{x} \in E^{(k)}$, then by $|\mathbf{x}|$ we shall mean

$$\left(\sum_{j=1}^{k} x_j^2 \right)^{1/2}.$$

Clearly, then $\mathbf{X}^{(n)} \to \mathbf{X}$ a.s. if and only if

$$|\mathbf{X}^{(n)} - \mathbf{X}| \to 0 \text{ a.s.}$$

Definition: Let $\{\mathbf{X}^{(n)}\}$ be a sequence of k-dimensional vector random variables. We say $\{\mathbf{X}^{(n)}\}$ converges in probability or is Cauchy convergent in probability (and we write $\mathbf{X}^{(n)} \overset{P}{\to}$) if for every $\epsilon > 0$,

$$P[\,|\mathbf{X}^{(n)} - \mathbf{X}^{(m)}| \geq \epsilon] \to 0$$

as $n, m \to \infty$. If $\mathbf{X}$ is a k-dimensional vector random variable, we say $\mathbf{X}^{(n)} \overset{P}{\to} \mathbf{X}$ if and only if

$$P[\,|\mathbf{X}^{(n)} - \mathbf{X}| \geq \epsilon] \to 0$$

as $n \to \infty$ for every $\epsilon > 0$.

Theorem 1. If $\mathbf{X}, \mathbf{X}^{(1)}, \mathbf{X}^{(2)}, \cdots$ are k-dimensional vector random variables, and if $\mathbf{X}^{(n)} \to \mathbf{X}$ a.s., then $\mathbf{X}^{(n)} \xrightarrow{P} \mathbf{X}$.

Proof: Let $\epsilon > 0$, $\zeta > 0$ be arbitrary. By Egorov's theorem there exists an event $A_\zeta \in \mathfrak{a}$ such that $P(A_\zeta) < \zeta$ and such that $\{\mathbf{X}^{(n)}\}$ converges *uniformly* over $A_\zeta{}^c$. Let N_ϵ be so large that $n > N_\epsilon$ implies that

$$| \mathbf{X}^{(n)}(\omega) - \mathbf{X}(\omega) | < \epsilon$$

for all $\omega \in A_\zeta{}^c$. Then for $n > N_\epsilon$, we have

$$P[\, | \mathbf{X}^{(n)} - \mathbf{X} | \geq \epsilon] < \zeta.$$

Q.E.D.

Theorem 2. If $\{\mathbf{X}^{(n)}\}$ is a sequence of k-dimensional random variables, and if $\mathbf{X}^{(n)} \xrightarrow{P}$, then there exists a subsequence $\{\mathbf{X}^{(n_k)}\}$ of $\{\mathbf{X}^{(n)}\}$ such that $\mathbf{X}^{(n_k)} \xrightarrow{\text{a.s.}}$ as $k \to \infty$.

Proof: Let $\epsilon_n > 0$ be such that $\sum \epsilon_n < \infty$. By hypothesis there exists an integer n_1 such that for all $n > n_1$ we have

$$P[\, | \mathbf{X}^{(n_1)} - \mathbf{X}^{(n)} | \geq \epsilon_1] < \epsilon_1.$$

In general, for $j \geq 2$, there exists an $n_j > n_{j-1}$ such that, for all $n > n_j$,

$$P[\, | \mathbf{X}^{(n)} - \mathbf{X}^{(n_j)} | \geq \epsilon_j] < \epsilon_j.$$

If we denote

$$A_j = [\, | \mathbf{X}^{(n_j)} - \mathbf{X}^{(n_{j+1})} | \geq \epsilon_j],$$

then we note that $\sum P(A_j) < \sum \epsilon_j < \infty$. By the Borel-Cantelli lemma, $P[A_n \text{ i.o.}] = 0$, or $P[A_n \text{ i.o.}]^c = 1$. Let $\epsilon > 0$ and let $\omega \in [A_n \text{ i.o.}]^c$. Then there is a j_0 such that for all $j > j_0$,

$$| \mathbf{X}^{(n_j)}(\omega) - \mathbf{X}^{(n_{j+1})}(\omega) | < \epsilon_j.$$

Let $j_1 \geq j_0$ be such that $\sum_{j=j_1}^{\infty} \epsilon_j < \epsilon$. Then for $r > j_1$ and $s > j_1$, and $r < s$ we have

$$| \mathbf{X}^{(n_r)}(\omega) - \mathbf{X}^{(n_s)}(\omega) | \leq \sum_{m=r}^{s-1} | \mathbf{X}^{(n_m)}(\omega) - \mathbf{X}^{(n_{m+1})}(\omega) |$$

$$< \sum_{m=r}^{s-1} \epsilon_m < \sum_{m=j_1}^{\infty} \epsilon_m < \epsilon,$$

which proves that $\mathbf{X}^{(n_j)} \xrightarrow{\text{a.s.}}$. Q.E.D.

The following lemma is used frequently from here on.

Lemma 1. If $\mathbf{X}$ and $\mathbf{Y}$ are k-dimensional vector random variables, then

$$P[\,|\,\mathbf{X} + \mathbf{Y}\,| \geq \epsilon] \leq P[\,|\,\mathbf{X}\,| \geq \tfrac{1}{2}\epsilon] + P[\,|\,\mathbf{Y}\,| \geq \tfrac{1}{2}\epsilon]$$

for every $\epsilon > 0$.

Proof: We first verify that

$$[\,|\,\mathbf{X} + \mathbf{Y}\,| \geq \epsilon] \subset [\,|\,\mathbf{X}\,| \geq \tfrac{1}{2}\epsilon] \cup [\,|\,\mathbf{Y}\,| \geq \tfrac{1}{2}\epsilon].$$

If $|\,\mathbf{X}(\omega) + \mathbf{Y}(\omega)\,| \geq \epsilon$, then because of the triangle inequality,

$$|\,\mathbf{X}(\omega)\,| + |\,\mathbf{Y}(\omega)\,| \geq \epsilon.$$

Now if the sum of two numbers is equal to or greater than ϵ, then at least one of the numbers must be as large or larger than $\tfrac{1}{2}\epsilon$; that is,

$$|\,\mathbf{X}(\omega)\,| \geq \tfrac{1}{2}\epsilon \quad \text{or} \quad |\,\mathbf{Y}(\omega)\,| \geq \tfrac{1}{2}\epsilon.$$

Thus we have proved the above relation between events. Taking probabilities of both sides and applying Boole's inequality, we obtain the conclusion.

Theorem 3. If $\{\mathbf{X}^{(n)}\}$ is a sequence of k-dimensional vector random variables, then $\{\mathbf{X}^{(n)}\}$ converges in probability if and only if there exists a k-dimensional vector random variable $\mathbf{X}$ such that $\mathbf{X}^{(n)} \xrightarrow{P} \mathbf{X}$. In this case $\mathbf{X}$ is unique except over an event of probability zero.

Proof: The uniqueness of $\mathbf{X}$ is easily proved. Suppose $\mathbf{X}^{(n)} \xrightarrow{P} \mathbf{X}$, and suppose $\mathbf{X}^{(n)} \xrightarrow{P} \mathbf{Y}$. Then by Lemma 1, for every $\epsilon > 0$,
$0 \leq P[\,|\,\mathbf{X} - \mathbf{Y}\,| \geq \epsilon]$

$$\leq P[\,|\,\mathbf{X} - \mathbf{X}^{(n)}\,| \geq \tfrac{1}{2}\epsilon] + P[\,|\,\mathbf{X}^{(n)} - \mathbf{Y}\,| \geq \tfrac{1}{2}\epsilon] \to 0$$

as $n \to \infty$, and thus $P[\,|\,\mathbf{X} - \mathbf{Y}\,| \geq \epsilon] = 0$. Thus $P[\mathbf{X} = \mathbf{Y}] = 1$. Now suppose $\mathbf{X}^{(n)} \xrightarrow{P}$. Let $\{\epsilon_j\}$ and $\{n_j\}$ be as in the proof of Theorem 2, and let $\mathbf{X}$ be such that $\mathbf{X}^{(n_k)} \xrightarrow{\text{a.s.}} \mathbf{X}$. By Lemma 1, for every $\epsilon > 0$,

$$P[\,|\,\mathbf{X}^{(n)} - \mathbf{X}\,| \geq \epsilon] \leq P[\,|\,\mathbf{X}^{(n)} - \mathbf{X}^{(n_k)}\,| \geq \tfrac{1}{2}\epsilon]$$

$$+ P[\,|\,\mathbf{X}^{(n_k)} - \mathbf{X}\,| \geq \tfrac{1}{2}\epsilon].$$

By Theorem 1,

$$P[\,|\,\mathbf{X}^{(n_k)} - \mathbf{X}\,|\, \geq \tfrac{1}{2}\epsilon] \to 0,$$

and since $\mathbf{X}^{(n)} \xrightarrow{P}$, the first term on the right converges to zero as $k \to \infty$ and $n \to \infty$. Thus $\mathbf{X}^{(n)} \xrightarrow{P} \mathbf{X}$.

Conversely, if $\mathbf{X}^{(n)} \xrightarrow{P} \mathbf{X}$, we have (by Lemma 1)

$$P[\,|\,\mathbf{X}^{(n)} - \mathbf{X}^{(m)}\,|\, \geq \epsilon] \leq P[\,|\,\mathbf{X}^{(n)} - \mathbf{X}\,|\, \geq \tfrac{1}{2}\epsilon]$$
$$+ P[\,|\,\mathbf{X} - \mathbf{X}^{(m)}\,|\, \geq \tfrac{1}{2}\epsilon],$$

and both terms on the right tend to zero as $n,\ m \to \infty$, thus proving $\mathbf{X}^{(n)} \xrightarrow{P}$. Q.E.D.

A useful theorem in probability and statistics is one which states, in a particular case, that if $\{X_n\}$ and $\{Y_n\}$ are two sequences of random variables, if $X_n \xrightarrow{P} X$, if $Y_n \xrightarrow{P} Y$, and if f is a continuous function over $E^{(2)}$, then $f(X_n, Y_n) \xrightarrow{P} f(X, Y)$. The aim of the next two theorems and their corollaries is to establish a general form of such a result.

Theorem 4. Let $\mathbf{X}, \mathbf{X}^{(1)}, \mathbf{X}^{(2)}, \cdots$ be a sequence of k-dimensional vector random variables. Then a necessary and sufficient condition that $\mathbf{X}^{(n)} \xrightarrow{P} \mathbf{X}$ is that every subsequence of $\{\mathbf{X}^{(n)}\}$ contains a subsequence which converges almost surely to $\mathbf{X}$.

Proof: If $\mathbf{X}^{(n)} \xrightarrow{P} \mathbf{X}$, then by the definition of convergence in probability, every subsequence $\{\mathbf{X}^{(n_k)}\}$ of $\{\mathbf{X}^{(n)}\}$ converges in probability to $\mathbf{X}$. Every such subsequence, because of Theorem 2, contains a subsequence which converges almost surely to $\mathbf{X}$. Thus the condition is necessary. In order to prove sufficiency, assume that it is not true that $\mathbf{X}^{(n)} \xrightarrow{P} \mathbf{X}$. Then there exists an $\epsilon_0 > 0$ and a $\delta_0 > 0$ such that

$$\lim \sup P[\,|\,\mathbf{X}^{(n)} - \mathbf{X}\,|\, \geq \epsilon_0] = \delta_0 > 0.$$

Hence there exists a subsequence $\{\mathbf{X}^{(n_k)}\}$ of $\{\mathbf{X}^{(n)}\}$ such that

$$P[\,|\,\mathbf{X}^{(n_k)} - \mathbf{X}\,|\, \geq \epsilon_0] \to \delta_0 > 0 \qquad \text{as} \quad k \to \infty.$$

Therefore, neither $\{\mathbf{X}^{(n_k)}\}$ nor any of its subsequences converges in probability to $\mathbf{X}$. Hence no subsequence of $\{\mathbf{X}^{(n_k)}\}$ converges almost surely to $\mathbf{X}$, for if a subsequence did, it would also converge in probability to $\mathbf{X}$, thus violating the fact that no subsequence of $\{\mathbf{X}^{(n_k)}\}$ does converge in probability to $\mathbf{X}$. Thus we have a contradiction. Q.E.D.

Theorem 5. If $\mathbf{X}$, $\mathbf{X}^{(1)}$, $\mathbf{X}^{(2)}$, $\cdots$ is a sequence of k-dimensional random variables, if $f: E^{(k)} \to E^{(m)}$ is a measurable mapping which is continuous over a Borel set $B \subset E^{(k)}$ for which $P[\mathbf{X} \in B] = 1$, and if $\mathbf{X}^{(n)} \overset{P}{\to} \mathbf{X}$, then $f(\mathbf{X}^{(n)}) \overset{P}{\to} f(\mathbf{X})$.

Proof: Let $\{ f(\mathbf{X}^{(n_j)}) \}$ be any subsequence of $\{ f(\mathbf{X}^{(n)}) \}$. By Theorem 4 we need only show that there is a subsequence $\{ f(\mathbf{X}^{(r_j)}) \}$ of $\{ f(\mathbf{X}^{(n_j)}) \}$ which converges almost surely. Since $\mathbf{X}^{(n_j)} \overset{P}{\to} \mathbf{X}$, then by Theorem 2 there is a subsequence $\{\mathbf{X}^{(r_j)}\}$ of $\{\mathbf{X}^{(n_j)}\}$ such that $\mathbf{X}^{(r_j)} \overset{\text{a.s.}}{\to} \mathbf{X}$. Let

$$A = [\mathbf{X}^{(r_j)} \to \mathbf{X}][\mathbf{X} \in B].$$

Clearly $P(A) = 1$, and $\omega \in A$ implies that $\mathbf{X}^{(r_j)}(\omega) \to \mathbf{X}(\omega) \in B$. Since f is continuous over B, then

$$f(\mathbf{X}^{(r_j)}(\omega)) \to f(\mathbf{X}(\omega))$$

as $j \to \infty$ for all $\omega \in A$; that is,

$$f(\mathbf{X}^{(r_j)}) \to f(\mathbf{X}) \text{ a.s.}$$

Hence we have proved that $f(\mathbf{X}^{(n)}) \overset{P}{\to} f(\mathbf{X})$.

Corollary 1. If X, X_1, X_2, $\cdots$ and Y, Y_1, Y_2, $\cdots$ are two sequences of random variables, if $X_n \overset{P}{\to} X$ and $Y_n \overset{P}{\to} Y$, and if f is a measurable function defined over $E^{(2)}$ such that $P[(X, Y) \in Cont\ f] = 1$, then $f(X_n, Y_n) \overset{P}{\to} f(X, Y)$.

Proof: We must first show that $X_n \overset{P}{\to} X$ and $Y_n \overset{P}{\to} Y$ imply that $(X_n, Y_n) \overset{P}{\to} (X, Y)$. If 0 denotes a random variable which equals zero with probability one, then

$$| (X_n, Y_n) - (X, Y) | \le |(X_n, 0) - (X, 0) | + | (0, Y_n) - (0, Y) |.$$

Since

$$| (X_n, 0) - (X, 0) | = | X_n - X |$$

and

$$| (0, Y_n) - (0, Y) | = | Y_n - Y |,$$

we obtain the fact that $(X_n, Y_n) \overset{P}{\to} (X, Y)$ by Lemma 1. The conclusion of the corollary now follows immediately from Theorem 5.

Corollary 2. If $\{X_n\}$, $\{Y_n\}$, X and Y satisfy the hypothesis of Corollary 1, then

(a) $aX_n + bY_n \xrightarrow{P} aX + bY$, where a, b are constants,

(b) $1/X_n \xrightarrow{P} 1/X$ provided that $P[X_n \neq 0] = P[X \neq 0] = 1$ for all n, and

(c) $X_n Y_n \xrightarrow{P} XY$.

Proof: These conclusions follow immediately from Theorem 5 and Corollary 1.

The following theorem is also a corollary to Theorem 5.

Theorem 6. If $\mathbf{X}$, $\mathbf{X}^{(1)}$, $\mathbf{X}^{(2)}$, $\cdots$ is a sequence of k-dimensional random variables, and if $\mathbf{X}^{(n)} \xrightarrow{P} \mathbf{X}$, then $\mathcal{L}(\mathbf{X}^{(n)}) \to \mathcal{L}(\mathbf{X})$.

Proof: For fixed $\mathbf{u} \in E^{(k)}$, $\exp\left[i(\mathbf{u}, \mathbf{x})\right]$ is a continuous mapping of $E^{(k)}$ into $E^{(2)}$ (as a function of $\mathbf{x}$). Hence by Theorem 5, $\exp i(\mathbf{u}, \mathbf{X}^{(n)}) \xrightarrow{P} \exp i(\mathbf{u}, \mathbf{X})$. By the Lebesgue dominated convergence theorem (for convergence in measure), we have

$$E \exp i(\mathbf{u}, \mathbf{X}^{(n)}) \to E \exp i(\mathbf{u}, \mathbf{X}) \qquad \text{as} \quad n \to \infty.$$

The continuity theorem implies that $\mathcal{L}(\mathbf{X}^{(n)}) \to \mathcal{L}(\mathbf{X})$.

Corollary. If $\mathbf{X}$ and $\{\mathbf{X}^{(n)}\}$ are k-dimensional vector random variables, and if $\mathbf{X}^{(n)} \to \mathbf{X}$ a.s., then $\mathcal{L}(\mathbf{X}^{(n)}) \to \mathcal{L}(\mathbf{X})$.

Proof: This is an immediate consequence of Theorem 1 and Theorem 6.

It should be noted that the two corollaries to Theorem 5 and Theorem 6 make strong use of the general assumptions of Theorem 5.

Another type of convergence that is frequently encountered is convergence in rth mean.

Definition: If $\{X_n\}$ is a sequence of random variables, it is said to converge in rth mean, for some $r \geq 1$, if $E \mid X_n \mid^r < \infty$ for all n and

$$E \mid X_n - X_m \mid^r \to 0$$

as $n \to \infty$ and $m \to \infty$. We denote this by $X_n \xrightarrow{r}$. If X is a random variable, we say $\{X_n\}$ converges in rth mean to X if $E \mid X \mid^r < \infty$ and

$$E \mid X_n - X \mid^r \to 0 \qquad \text{as} \quad n \to \infty.$$

In this case we write $X_n \xrightarrow{r} X$.

The basic properties of convergence in rth mean are included in the exercises.

EXERCISES

1. Let $X, X_1, X_2, \cdots$ be random variables. Then $X_n \xrightarrow{P} X$ if and only if

$$E\left(\frac{\mid X_n - X \mid}{1 + \mid X_n - X \mid}\right) \to 0 \qquad \text{as} \quad n \to \infty.$$

(Hint: use Theorem 2 in Section 2.4 and the Lebesgue dominated convergence theorem.)

(In all the exercises that follow, $X, X_1, X_2, \cdots$ and $Y, Y_1, Y_2, \cdots$ are random variables over the same probability space.)

2. Prove: If $X_n \xrightarrow{r}$, then $X_n \xrightarrow{P}$, and if $X_n \xrightarrow{r} X$, then $X_n \xrightarrow{P} X$.

3. Prove: If $X_n \xrightarrow{r}$, there exists a subsequence $\{X_{n_k}\}$ of $\{X_n\}$ such that $X_{n_k} \xrightarrow{\text{a.s.}}$.

4. Prove: If $X_n \xrightarrow{r}$, then there exists a random variable X such that $E \mid X \mid^r < \infty$ and $X_n \xrightarrow{r} X$, and conversely.

5. Prove: If $1 \leq r' < r$, and if $X_n \xrightarrow{r} X$, then $X_n \xrightarrow{r'} X$.

6. Prove: If $\mathcal{L}(X_n) \to \mathcal{L}(X)$, and if $P[X = 0] = 1$, then $X_n \xrightarrow{P} X$.

7. Assume F_X and $\{F_{X_n}\}$ are all absolutely continuous with densities f and $\{f_n\}$. If $f_n \to f$ a.e. with respect to Lebesgue measure, then $\mathcal{L}(X_n) \to \mathcal{L}(X)$.

8. Prove: $X_n \xrightarrow{P} X$ and $Y_n \xrightarrow{P} Y$ if and only if $(X_n, Y_n) \xrightarrow{P} (X, Y)$.

9. Construct examples over the unit-interval probability space to show that neither of convergence in rth mean and convergence almost surely implies the other.

CHAPTER 5

Strong Limit Theorems for Independent Random Variables

═══

5.1. Almost Sure Convergence of Series of Independent Random Variables

The strong limit theorems (or strong limit laws) in probability refer to those theorems which deal with almost sure convergence of a sequence of random variables. This chapter contains some of the best known and most widely used theorems of this type. In this section we are concerned with almost sure convergence of series of independent random variables.

If $\{X_n\}$ is a sequence of random variables, we say that the series $\sum X_n$ converges almost surely if the sequence of partial sums $\{S_n\}$ converges almost surely, where $S_n = X_1 + \cdots + X_n$. We also say that $\sum X_n$ converges in law if there is a distribution function F such that $F_{S_n} \xrightarrow{c} F$. Also, $\sum X_n$ converges in probability (or in rth mean) if there exists a random variable X such that $S_n \xrightarrow{P} X$ (or $S_n \xrightarrow{r} X$). A valuable tool for the strong limit theorems is the following.

Theorem 1 (Kolmogorov's Inequalities). If X_1, X_2, $\cdots$, X_n are independent random variables, each with finite second moment, then

$$P[\max_{1 \leq k \leq n} | S_k - ES_k | \geq \epsilon] \leq \frac{1}{\epsilon^2} \sum_{j=1}^{n} \operatorname{Var} X_j$$

for every $\epsilon > 0$. If in addition there is a constant K such that

$$P[\,|\,X_k\,|\,\leq K\,] = 1, \qquad 1 \leq k \leq n,$$

then

$$P[\,\max_{1\leq k\leq n}|\,S_k - ES_k\,|\,\geq \epsilon\,]$$

$$\geq 1 - (\epsilon + K)^2/\{\sum_{k=1}^{n} \mathrm{Var}\,X_k + (\epsilon + K)^2 - \epsilon^2\}.$$

Proof: We may assume without loss of generality that $EX_k = 0$ for all k. Let us denote

$$A_k = [\,\max_{1\leq j\leq k}|\,S_j\,|\,<\epsilon\,] = \bigcap_{j=1}^{k}[\,|\,S_j\,|\,<\epsilon\,],$$

and let

$$B_k = A_{k-1}A_k{}^c = (\bigcap_{j=1}^{k-1}[\,|\,S_j\,|\,<\epsilon\,])[\,|\,S_k\,|\,\geq \epsilon\,],$$

that is, B_k is the event that S_k is the first partial sum whose absolute value is $\geq \epsilon$. Let $A_0 = \Omega$. It is easy to verify that

$$A_n{}^c = [\,\max_{1\leq j\leq n}|\,S_j\,|\,\geq \epsilon\,] = \bigcup_{k=1}^{n} B_k;$$

further, $B_1, \cdots, B_n$ are disjoint events. For $1 \leq k \leq n$,

$$\int_{B_k} S_n{}^2\,dP = E(S_n I_{B_k})^2$$

$$= E((S_n - S_k)I_{B_k} + S_k I_{B_k})^2$$

$$= E((S_n - S_k)I_{B_k})^2 + 2E((S_n - S_k)S_k I_{B_k}) + E(S_k I_{B_k})^2.$$

Since $S_n - S_k$ and $S_k I_{B_k}$ are independent, then by Theorem 1 of Section 3.2 and by the fact that $EX_j = 0$ for $1 \leq j \leq n$ we have

$$E((S_n - S_k)S_k I_{B_k}) = E(S_n - S_k)E(S_k I_{B_k}) = 0.$$

Thus

$$\int_{B_k} S_n{}^2\,dP \geq E(S_k I_{B_k})^2 \geq \epsilon^2 P(B_k).$$

From this we obtain

$$\sum_{k=1}^{n} \text{Var}\,(X_k) \;=\; \text{Var}\,S_n \;\geq\; \int_{A_n{}^c} S_n{}^2\,dP \;=\; \sum_{k=1}^{n} \int_{B_k} S_n{}^2\,dP \;\geq\; \epsilon^2 P\Big(\overset{n}{\underset{k=1}{\cup}} B_k\Big),$$

from which we obtain the first inequality of the theorem. In order to obtain the second inequality of the theorem we first observe that A_k and B_k are disjoint and $A_k \cup B_k = A_{k-1}$. Hence

$$S_{k-1}I_{A_{k-1}} + X_k I_{A_{k-1}} = S_k I_{A_{k-1}} = S_k I_{A_k} + S_k I_{B_k}.$$

Squaring and taking expectations we obtain

$$E(S_{k-1}I_{A_{k-1}})^2 + 2E(S_{k-1}X_k I_{A_{k-1}}) + E(X_k{}^2 I_{A_{k-1}})$$

$$= E(S_k I_{A_k})^2 + 2E(S_k{}^2 I_{A_k} I_{B_k}) + E(S_k I_{B_k})^2.$$

The random variables $S_{k-1}I_{A_{k-1}}$ and X_k are independent, and $EX_k = 0$. Thus by Theorem 1 in Section 3.2 the middle term of the left hand side above vanishes. Since $A_k B_k = \phi$, then $I_{A_k} I_{B_k} = 0$, and hence

$$E(S_k{}^2 I_{A_k} I_{B_k}) = 0.$$

Also, X_k and $I_{A_{k-1}}$ are independent, so

$$E(X_k{}^2 I_{A_{k-1}}) = (\text{Var}\,X_k) P(A_{k-1}).$$

Thus the above equation becomes

$$E(S_{k-1}I_{A_{k-1}})^2 + (\text{Var}\,X_k) P(A_{k-1}) = E(S_k I_{A_k})^2 + E(S_k I_{B_k})^2.$$

Since $A_k \subset A_{k-1}$, we have $P(A_{k-1}) \geq P(A_k)$. Also, $P[\,|\,X_k\,| \leq K\,] = 1$. Hence over an event of probability one,

$$|\,S_k I_{B_k}\,| \leq |\,S_{k-1}I_{B_k}\,| + |\,X_k I_{B_k}\,| \leq (\epsilon + K) I_{B_k}.$$

Applying this inequality and the fact that $P(A_{k-1}) \geq P(A_n)$ to the last equation we get

$$E(S_{k-1}I_{A_{k-1}})^2 + P(A_n)\,\text{Var}\,(X_k) \leq E(S_k I_{A_k})^2 + (\epsilon + K)^2 P(B_k),$$

or

$$E(S_k I_{A_k})^2 - E(S_{k-1}I_{A_{k-1}})^2 \geq P(A_n)\,\text{Var}\,X_k - (\epsilon + K)^2 P(B_k).$$

Summing both sides from $k = 1$ to $k = n$, and remembering that $S_0 = 0$,

we obtain

$$E(S_n I_{A_n})^2 \geq \sum_{k=1}^{n} \text{Var } (X_k) - P(A_n{}^c)(\sum_{k=1}^{n} \text{Var } (X_k) + (\epsilon + K)^2).$$

Since $E(S_n I_{A_n})^2 \leq \epsilon^2 P(A_n)$, we have

$$\epsilon^2 + P(A_n{}^c)(\sum_{k=1}^{n} \text{Var } (X_k) + (\epsilon + K)^2 - \epsilon^2)$$

$$\geq \sum_{k=1}^{n} \text{Var } (X_k),$$

or

$$P(A_n{}^c) \geq \frac{\sum_{k=1}^{n} \text{Var } (X_k) - \epsilon^2}{\sum_{k=1}^{n} \text{Var } (X_k) + (\epsilon + K)^2 - \epsilon^2}$$

$$= 1 - \frac{(\epsilon + K)^2}{\sum_{k=1}^{n} \text{Var } (X_k) + (\epsilon + K)^2 - \epsilon^2}.$$

This completes the proof.

Theorem 2. Let $\{X_n\}$ be a sequence of independent random variables with finite second moments. If

$$\sum_{k=1}^{\infty} \text{Var } (X_k) < \infty,$$

then $\sum_{n=1}^{\infty} (X_n - EX_n)$ converges almost surely.

Proof: We denote $S_n = \sum_{j=1}^{n}(X_j - EX_j)$. We first prove that $\limsup S_n$ and $\liminf S_n$ are finite almost surely. Indeed, for every $m \geq 1$,

$$\limsup_{n\to\infty} | S_n | \leq \limsup_{n\to\infty} | S_n - S_m | + | S_m |$$

$$\leq \sup_{n\geq m} | S_n - S_m | + | S_m |.$$

By Kolmogorov's inequality (Theorem 1) we have

$$P[\sup_{n \geq m} | S_n - S_m | \geq 2\epsilon] \leq P(\bigcup_{n=m}^{\infty} [| S_n - S_m | \geq \epsilon])$$

$$= \lim_{M \to \infty} P(\bigcup_{n=m}^{M} [| S_n - S_m | \geq \epsilon])$$

$$= \lim_{M \to \infty} P[\max_{m \leq n \leq M} | S_n - S_m | \geq \epsilon]$$

$$\leq \frac{1}{\epsilon^2} \sum_{n=m}^{\infty} \text{Var} (X_k) < \infty,$$

the finiteness of this last term being implied by hypothesis. Let $\epsilon \to \infty$, and the first probability tends to zero. Thus we have shown that lim sup S_n and likewise lim inf S_n are finite almost surely. In order to prove $S_n \overset{\text{a.s.}}{\to}$ we need only show that $P[\text{lim sup } S_n - \text{lim inf } S_n \geq \epsilon] = 0$ for every $\epsilon > 0$. In order to do this we first observe that if $\{a_n\}$ is a set of numbers, then

$$\sup | a_n | = \max \{ | \sup a_n |, | \inf a_n | \},$$

and $| \sup a_n | > \epsilon$ or $| \inf a_n | > \epsilon$ implies $\sup | a_n | > \epsilon$. Thus for every positive integer m and every $\epsilon > 0$,

$[\text{lim sup } S_n - \text{lim inf } S_n \geq 4\epsilon]$

$$\subset [\sup_{n \geq m} S_n - \inf_{n \geq m} S_n \geq 4\epsilon]$$

$$\subset [| \sup_{n \geq m} S_n - S_m | \geq 2\epsilon] \cup [| S_m - \inf_{n \geq m} S_n | \geq 2\epsilon]$$

$$\subset [\sup_{n \geq m} | S_n - S_m | \geq 2\epsilon].$$

By Kolmogorov's inequality, then

$P[\text{lim sup } S_n - \text{lim inf } S_n \geq 4\epsilon]$

$$\leq \frac{1}{\epsilon^2} \sum_{j=m}^{\infty} \text{Var} (X_j) \to 0 \qquad \text{as} \quad m \to \infty,$$

which proves the theorem.

Corollary. If $\{X_n\}$ is a sequence of independent random variables with finite second moments, if $\sum_{n=1}^{\infty} \text{Var}(X_n) < \infty$, and if $\sum_{n=1}^{\infty} E(X_n)$ converges, then $\sum_{n=1}^{\infty} X_n$ converges almost surely.

Proof: This corollary is an immediate consequence of Theorem 2 and the fact that the sum of two convergent series is convergent.

If the random variables $\{X_n\}$ are uniformly bounded by a constant, then we are able to show in the following theorem that the converse to the above corollary is true.

Theorem 3. Let $\{X_n\}$ be a sequence of independent random variables such that $P[\,|X_n| \leq K\,] = 1$ for all n and some $K > 0$. If $\sum_{n=1}^{\infty} X_n$ converges almost surely, then $\sum_{n=1}^{\infty} \text{Var}(X_n) < \infty$ and $\sum_{n=1}^{\infty} E(X_n)$ converges.

Proof: Let $\{X_n'\}$ be a sequence of random variables such that $F_{X_n} = F_{X_n'}$ for all n and such that $\{X_1, X_1', X_2, X_2', \cdots\}$ are all independent. (The Kolmogorov-Daniell theorem guarantees that such a sequence exists.) Let us denote $X_n^s = X_n - X_n'$. Then $|X_n^s| \leq |X_n| + |X_n'| \leq 2K$ almost surely and $EX_n^s = 0$ for all n. Since $\sum_{n=1}^{\infty} X_n$ converges almost surely, then so does $\sum_{n=1}^{\infty} X_n'$ and $\sum_{n=1}^{\infty} X_n^s$. (This follows from the discussion following the proof of the Kolmogorov-Daniell theorem.) Now

$$\Big[\sum_{n=1}^{\infty} X_n^s \text{ diverges}\Big] = \bigcup_{N=1}^{\infty} \bigcap_{n=1}^{\infty} \bigcup_{m=n}^{\infty} [\sup_{k \geq m} |S_k - S_m| \geq N^{-1}].$$

We shall prove that $\sum_{n=1}^{\infty} \text{Var}(X_n^s) < \infty$. Let us assume the contrary. Then by the second inequality in Theorem 1, if we denote

$$S_n^s = X_1^s + \cdots + X_n^s,$$

then

$$P[\max_{m \leq k \leq r} |S_k^s - S_m^s| \geq N^{-1}]$$

$$= P\Big(\bigcup_{k=m}^{n} [\,|S_k^s - S_m^s| \geq N^{-1}]\Big)$$

$$\geq 1 - \frac{(1/N + 2K)^2}{\sum_{k=m+1}^{r} \text{Var } X_k^s + (1/N + 2K)^2 - (1/N)^2}.$$

Hence for every positive integer m,

$$P[\sup_{k \geq m} | S_k^s - S_m^s | \geq N^{-1}]$$

$$\geq P(\bigcup_{r=m}^{\infty} [\max_{r \geq k \geq m} | S_k^s - S_m^s | \geq N^{-1}])$$

$$= \lim_{r \to \infty} P[\max_{r \geq k \geq m} | S_k^s - S_m^s | \geq N^{-1}]$$

$$\geq \lim_{r \to \infty} \left\{ 1 - \frac{((1/N) + 2K)^2}{\displaystyle\sum_{k=m+1}^{r} \text{Var} (X_k^s) + (4K/N) + 4K^2} \right\} = 1.$$

From this it follows that

$$P[\sum_{n=1}^{\infty} X_n^s \text{ diverges}] = 1,$$

which contradicts the fact pointed out above that $\sum_{n=1}^{\infty} X_n^s$ converges almost surely. Thus $\sum_{n=1}^{\infty} \text{Var} (X_n^s) < \infty$, as we wished to prove. Since $\text{Var} (X_n^s) = 2 \text{ Var} (X_n)$, we have $\sum_{n=1}^{\infty} \text{Var} (X_n) < \infty$. By Theorem 2 this implies that $\sum_{n=1}^{\infty} (X_n - EX_n)$ converges almost surely. By hypothesis, $\sum_{n=1}^{\infty} X_n$ converges almost surely, and hence $\sum_{n=1}^{\infty} EX_n$ converges. Q.E.D.

The most useful criterion for establishing almost sure convergence of a series of independent random variables where no assumptions are made on the existence of moments is the following theorem. Notice in the proof how much use is made of the method of truncation which was discussed just after the Borel-Cantelli lemma.

Theorem 4 (Kolmogorov's Three Series Theorem). Let $\{X_n\}$ be a sequence of independent random variables. The series $\sum_{n=1}^{\infty} X_n$ converges almost surely if and only if for some constant $c > 0$ the following three series converge:

(a) $\displaystyle\sum_{n=1}^{\infty} P[| X_n | \geq c]$

(b) $\displaystyle\sum_{n=1}^{\infty} E(X_n I_{[|X_n| < c]})$

(c) $\displaystyle\sum_{n=1}^{\infty} \text{Var} (X_n I_{[|X_n| < c]}).$

Further, if the series (a), (b), (c) converge for one value of $c > 0$, then they converge for all $c > 0$.

Proof: Let us assume that the three series converge, and let

$$Y_n = X_n I_{[|X_n| < c]}.$$

By the corollary to Theorem 2, convergence of series (b) and (c) implies that $\sum_{n=1}^{\infty} Y_n$ converges almost surely. Convergence of (a) implies by the Borel-Cantelli lemma that $P[X_n \neq Y_n \text{ i.o.}] = 0$. Hence $\sum_{n=1}^{\infty} X_n$ converges almost surely. Conversely, if $\sum_{n=1}^{\infty} X_n$ converges almost surely, then $X_n \overset{\text{a.s.}}{\to} 0$. Hence $P(\lim \sup_{n \to \infty} [\,|X_n| \geq c]) = 0$ for every $c > 0$. By the Borel lemmas (Theorem 2 in Section 3.3), $\sum_{n=1}^{\infty} P[\,|X_n| \geq c]$ converges for every $c > 0$. By hypothesis and Corollary 2 to the Borel-Cantelli lemma, $\sum X_n I_{[|X_n| < c]}$ converges almost surely for every $c > 0$. Theorem 3 now implies that series (b) and (c) converge.

Exercises

1. Prove: If $\{X_n\}$ are random variables, then

$$\Big[\sum_{n=1}^{\infty} X_n \text{ diverges}\Big] = \bigcup_{N=1}^{\infty} \bigcap_{n=1}^{\infty} \bigcup_{m=n}^{\infty} [\sup_{k \geq m} |X_m + \cdots + X_k| \geq N^{-1}].$$

2. Prove that if the three series in Kolmogorov's three series theorem converge for one value of $c > 0$ they converge for all $c > 0$.

3. In the proof of Theorem 1 prove that $S_n - S_k$ and $S_k I_{B_k}$ are independent for $1 \leq k \leq n$.

4. Let $\{X_n\}$ be a sequence of independent random variables which satisfy $P[X_n = 1] = P[X_n = -1] = \frac{1}{2}$ for all n, and let $\{a_n\}$ be a sequence of real numbers. Prove that $\sum_{n=1}^{\infty} a_n X_n$ converges almost surely if and only if $\sum_{n=1}^{\infty} a_n^2 < \infty$.

5. Let $R_n(x)$ denote the nth Rademacher function over $[0, 1]$ (see Problem 2 in Section 10). Show that if $\{b_n\}$ is any sequence of real numbers, $\sum_{n=1}^{\infty} b_n R_n$ converges almost everywhere (Lebesgue measure) if and only if $\sum_{n=1}^{\infty} b_n^2 < \infty$.

6. In Problem 4 let $a_n = (\frac{1}{2})^n$, and prove in two ways that the distribution function of $\sum_{n=1}^{\infty} a_n X_n$ is uniform over $[-1, 1]$, that is, it has a density $f(x) = \frac{1}{2}$ if $|x| \leq 1$ and $f(x) = 0$ if $|x| > 1$.

7. Let $\{X_n\}$ be a sequence of independent random variables such that $P[X_n = 0] = P[X_n = 1] = \frac{1}{2}$. Prove that

$$\sum_{n=1}^{\infty} \frac{2}{3} \left(\frac{1}{3}\right)^{n-1} X_n$$

converges almost surely and that its distribution function is the Cantor distribution defined in Section 2.1.

8. In Problem 4, let $a_n = \beta^n$, $0 < \beta < \frac{1}{2}$. Prove that the distribution function of $\sum_{n=1}^{\infty} a_n X_n$ is continuous singular.

9. Prove: If X is a random variable with a symmetric distribution function, then so is $XI_{[|X| \leq K]}$ for every $K > 0$, and $E(XI_{[|X| \leq K]}) = 0$.

5.2. Proof that Convergence in Law of a Series of Independent Random Variables Implies Almost Sure Convergence

A major theorem in probability theory is a theorem which states that if a series of independent random variables converges in law, then it converges almost surely. The major tools for proving this theorem are the Kolmogorov three series theorem proved in Section 5.1 and a technique of centering at medians.

Definition: If X is a random variable, then by a median of X we mean a *number* μX which satisfies

$$P[X \leq \mu X] \geq \frac{1}{2} \quad \text{and} \quad P[X \geq \mu X] \geq \frac{1}{2}.$$

A median of a distribution function F is a number μ such that $F(\mu) \geq \frac{1}{2}$ and $1 - F(\mu - 0) \geq \frac{1}{2}$.

It should be noted that median of a random variable or distribution function is not uniquely determined. For example if

$$F(x) = \begin{cases} 0 & \text{if } x < 0 \\ \frac{1}{2} & \text{if } 0 \leq x < 1 \\ 1 & \text{if } x \geq 1, \end{cases}$$

then any number in $[0, 1]$ is a median of F. (Note the use made of the indefinite article.) It should also be noted that a median of a random variable is a median of its distribution function, and a median of a distribution function is a median of any random variable whose distribution function it is.

Lemma 1. If X, Y are independent random variables with the same distribution function, then $2P[\,|\,X - Y\,| \geq c] \geq P[\,|\,X - \mu X\,| \geq c]$ for every $c > 0$ and every median μX of X.

Proof: We may take $\mu X = \mu Y$. Then

$$P[X - Y \geq c] = P[(X - \mu X) - (Y - \mu Y) \geq c]$$
$$\geq P([X - \mu X \geq c][Y - \mu Y \leq 0])$$
$$\geq \tfrac{1}{2}P[X - \mu X \geq c].$$

Similarly, $P[X - Y \leq -c] \geq \tfrac{1}{2}P[X - \mu X \leq -c]$, and, combining the two inequalities, we obtain the conclusion of the lemma.

Theorem 1. If $\{X_n\}$ is a sequence of independent random variables, and if $\sum_{n=1}^{\infty} X_n$ converges in law, then it converges almost surely.

Proof: Let $f_n(u)$ be the characteristic function of X_n. Then the sequence of partial products $\{\prod_{k=1}^{n} f_k(u)\}$ converges to a function $f(u)$ which is a characteristic function, and we write

$$f(u) = \prod_{k=1}^{\infty} f_k(u) = \lim_{n \to \infty} \prod_{k=1}^{n} f_k(u).$$

Let $\{X_n'\}$ be a sequence of random variables such that $F_{X_n} = F_{X_n'}$ for all n and such that $\{X_1, X_1', X_2, X_2', \cdots\}$ are all independent. (The Kolmogorov-Daniell theorem guarantees that this can be done.) Denote $X_n^{(s)} = X_n - X_n'$. It is easy to verify that the characteristic function of $X_n^{(s)}$ is $|\,f_n(u)\,|^2$, and hence by Proposition 5 in Section 2.5 the distribution function of $X_n^{(s)}$ is symmetric. We first prove that $\sum_{n=1}^{\infty} X_n^{(s)}$ converges almost surely.

Since $X_n^{(s)}I_{[|X_n^{(s)}|<K]}$ is symmetric and bounded, its expectation is zero, so the series $\sum_{n=1}^{\infty}E(X_n^{(s)}I_{[|X_n^{(s)}|<K]})$ converges. We next prove that the series $\sum_{n=1}^{\infty}\mathrm{Var}\,(X_n^{(s)}I_{[|X_n^{(s)}|<K]})$ converges. Since $\prod_{k=1}^{n}|f_k(u)|^2$ converges as $n\to\infty$ to $|f(u)|^2$, which is unequal to zero in a neighborhood of the origin, then the infinite product

$$\prod_{n=1}^{\infty}|f_n(u)|^2 = \prod_{n=1}^{\infty}(1-(1-|f_n(u)|^2))$$

converges in a neighborhood of 0; hence the series $\sum_{n=1}^{\infty}(1-|f_n(u)|^2)$ converges near 0. Let $F_n^{(s)}$ be the distribution function of X_n-X_n'. Then

$$1-|f_n(u)|^2 = \int_{-\infty}^{\infty}(1-\cos ux)\,dF_n^{(s)}(x)$$

$$\geq \int_{-K}^{K}(1-\cos ux)\,dF_n^{(s)}(x)$$

$$\geq \int_{-K}^{K}\left(\frac{u^2x^2}{2}-\frac{u^4x^4}{24}\right)dF_n^{(s)}(x)$$

$$\geq \frac{u^2}{2}\left(1-\frac{u^2K^2}{12}\right)\int_{-K}^{K}x^2\,dF_n^{(s)}(x)$$

$$\geq \frac{u^2}{2}\left(1-\frac{u^2K^2}{12}\right)\mathrm{Var}\,(X_n^{(s)}\,I_{[|X_n^{(s)}|<K]}).$$

Thus $\sum_{n=1}^{\infty}\mathrm{Var}\,(X_n^{(s)}\,I_{[|X_n^{(s)}|<K]})<\infty$ for every $K>0$. We now prove that $\sum_{n=1}^{\infty}P[\,|X_n^{(s)}|\geq K]<\infty$. We first note that in some neighborhood of zero,

$$-\sum_{n=1}^{\infty}\log|f_n(u)|^2 = -\log|f(u)|^2\geq 0,$$

or,

$$0\leq\sum_{n=1}^{\infty}(1-|f_n(u)|^2)\leq -\log|f(u)|^2.$$

Hence by Fubini's theorem we have in some neighborhood of 0, for $u > 0$,

$$\frac{1}{u} \int_0^u (1 - |f_n(v)|^2)\, dv = \frac{1}{u} \int_0^u \left(\int_{-\infty}^{\infty} (1 - \cos vx)\, dF_n^{(s)}(x) \right) dv$$

$$= \frac{1}{u} \int_{-\infty}^{\infty} \left(\int_0^u (1 - \cos vx)\, dv \right) dF_n^{(s)}(x)$$

$$= \int_{-\infty}^{\infty} \left(1 - \frac{\sin ux}{ux} \right) dF_n^{(s)}(x)$$

$$\geq \int_{|x| \geq 1/u} \left(1 - \frac{\sin ux}{ux} \right) dF_n^{(s)}(x)$$

$$\geq (1 - \sin 1) P[\,|X_n^{(s)}| \geq u^{-1}\,].$$

By the Lebesgue dominated convergence theorem, for u near 0, $u > 0$, we have

$$\infty > \frac{1}{u} \int_0^u (-\log |f(v)|^2)\, dv \geq \frac{1}{u} \int_0^u \left(\sum_{n=1}^{\infty} (1 - |f_n(v)|^2) \right) dv$$

$$\geq (1 - \sin 1) \sum_{n=1}^{\infty} P[\,|X_n| \geq u^{-1}\,].$$

If we take $K = 1/u$, then the three series of the three series theorem converge for at least one value of K, and hence by that theorem the series $\sum_{n=1}^{\infty} X_n^{(s)}$ converges almost surely.

We next prove that there exists a sequence of real numbers $\{a_n\}$ such that $\sum_{n=1}^{\infty} (X_n - a_n)$ converges almost surely. By Lemma 1 we have

$$\sum_{n=1}^{\infty} P[\,|X_n - \mu X_n| \geq K\,] \leq 2 \sum_{n=1}^{\infty} P[\,|X_n^{(s)}| \geq K\,] < \infty$$

for all $K > 0$. We prove at this point that

$$\sum_{n=1}^{\infty} \text{Var} \left((X_n - \mu X_n) I_{[|X_n - \mu X_n| \leq K]} \right) < \infty$$

for every $K > 0$. Indeed, if we take $\mu X_n' = \mu X_n$ and denote $U = X_n -$

μX_n and $V = X_n' - \mu X_n'$, then

$$| UI_{[|U|\leq K]} - VI_{[|V|\leq K]} |$$

$$\leq | (U - V)I_{[|U|\leq K][|V|\leq K]} |$$

$$+ | UI_{[|U|\leq K][|V|> K]} |$$

$$+ | VI_{[|V|\leq K][|U|> K]} |$$

$$\leq | U - V | I_{[|U-V|\leq 2K]} + K(I_{[|V|> K]} + I_{[|U|> K]}).$$

Squaring, taking expectations, and using the fact that

$$| U - V | I_{[|U-V|\leq 2K]} \leq 2K$$

almost surely, we obtain

$$E(UI_{[|U|\leq K]} - VI_{[|V|\leq K]})^2$$

$$\leq E((U - V)^2 I_{[|U-V|\leq 2K]})$$

$$+ K^2(P[\,|\,U\,|\,> K] + 2P[\,|\,U\,|\,> K]P[\,|\,V\,|\,> K]$$

$$+ P[\,|\,V\,|\,> K])$$

$$+ 4K^2(P[\,|\,U\,|\,> K] + P[\,|\,V\,|\,> K]).$$

Now $(U - V)I_{[|U-V|\leq 2K]}$ is a Borel-measurable function $h(\cdot, \cdot)$ of (U, V), and $h(u, v) = -h(v, u)$. Since U and V are independent and have the same distribution, the joint distribution of (U, V) is the same as that of (V, U). Hence

$$P[h(U, V) \in B] = P[(U, V) \in h^{-1}(B)]$$

$$= P[(V, U) \in h^{-1}(B)] = P[h(V, U) \in B]$$

$$= P[-h(U, V) \in B] = P[h(U, V) \in -B],$$

for every Borel set B, where $-B = \{-b \mid b \in B\}$. Hence the distribution function of $h(U, V)$ is symmetric and $Eh(U, V) = 0$. Thus

$$E(UI_{[|U|\leq K]} - VI_{[|V|\leq K]})^2 \leq \mathrm{Var}\,((U - V)I_{[|U-V|\leq 2K]})$$

$$+ 12K^2P[\,|\,U\,|\,> K].$$

The left hand side of this last inequality is easily seen to be equal to $2\,\mathrm{Var}\,(UI_{[|U|\leq K]})$. Since $\sum_{n=1}^{\infty} X_n^{(s)}$ converges almost surely, we have, by the above inequality, the Kolmogorov three series theorem, and Lemma 1 that

$$2\sum_{n=1}^{\infty} \text{Var}\left((X_n - \mu X_n)I_{[|X_n - \mu X_n| \leq K]}\right)$$

$$\leq \sum_{n=1}^{\infty} \text{Var}\left(X_n^{(s)}\, I_{[|X_n^{(s)}| \leq 2K]}\right)$$

$$+\, 24K^2 \sum_{n=1}^{\infty} P[\,|X_n^{(s)}| \geq K] < \infty.$$

This last inequality and Theorem 2 in Section 5.1 imply

$$\sum_{n=1}^{\infty} \left\{ (X_n - \mu X_n)I_{[|X_n - \mu X_n| \leq K]} - E((X_n - \mu X_n)I_{[|X_n - \mu X_n| \leq K]}) \right\}$$

converges almost surely. As we noted on page 118,

$$\sum_{n=1}^{\infty} P[\,|X_n - \mu X_n| > K] < \infty \qquad \text{for all} \quad K > 0,$$

so by the Borel-Cantelli lemma,

$$\sum_{n=1}^{\infty} \left\{ X_n - \mu X_n - E((X_n - \mu X_n)I_{[|X_n - \mu X_n| \leq K]}) \right\}$$

converges almost surely. Let

$$a_n = \mu X_n + E((X_n - \mu X_n)I_{[|X_n - \mu X_n| \leq K]}),$$

and we can conclude that $\sum_{n=1}^{\infty}(X_n - a_n)$ converges almost surely for a sequence $\{a_n\}$.

In order to conclude the proof that $\sum_{n=1}^{\infty} X_n$ converges almost surely we prove that $\sum_{n=1}^{\infty} a_n$ converges. Since $\sum_{k=1}^{n}(X_k - a_k)$ converges almost surely as $n \to \infty$ to some random variable X with characteristic function $g(u)$, we have $\prod_{k=1}^{n} \exp(-ia_k u)f_k(u) \to g(u)$ as $n \to \infty$ for all u. For all u in some neighborhood of 0, $g(u)\overline{f(u)} \neq 0$, so

$$\lim_{n \to \infty}\left(\prod_{k=1}^{n} \exp(-ia_k u)f_k(u)\right)\left(\prod_{j=1}^{n} \overline{f_j(u)}\right) \neq 0$$

or

$$\lim_{n \to \infty}\left\{\exp\left[-iu \sum_{j=1}^{n} a_j\right]\right\} \prod_{k=1}^{n} |f_k(u)|^2 \neq 0.$$

Since

$$\prod_{k=1}^{\infty} |f_k(u)|^2 \neq 0$$

in a neighborhood of 0, then $\lim_{n\to\infty} \exp\{-iu\sum_{k=1}^{n} a_k\}$ exists near 0. Let $s_n = a_1 + \cdots + a_n$. If $\{\exp(-ius_n)\}$ converges in $[-K, K]$, then it converges in $[-2K, 2K]$, since $\exp(-i2us_n) = [\exp(-ius_n)]^2$. Hence $\{\exp(-ius_n)\}$ converges for all u. Since

$$\lim_{n\to\infty} \exp(-ius_n) = \frac{g(u)\overline{f(u)}}{\displaystyle\prod_{k=1}^{\infty} |f_k(u)|^2}$$

near 0, since both numerator and denominator on the right side are 1 at $u = 0$ and are continuous at $u = 0$, then $\lim_{n\to\infty} \exp(-ius_n)$ is continuous at zero. Hence by the continuity theorem,

$$h(u) = \lim_{n\to\infty} \exp(-ius_n)$$

is a characteristic function, since each $\exp(-ius_n)$ *is* a characteristic function of a random variable W_n for which $P[W_n = -s_n] = 1$. Hence, if G_n denotes the distribution function of W_n, we know by the continuity theorem that there exists a distribution function G such that $G_n \xrightarrow{c} G$. Easily, there are no points $x \in Cont\ G$ such that $0 < G(x) < 1$. Hence, for some x_0, $G(x) = 0$ for all $x < x_0$ and $G(x) = 1$ for all $x \geq x_0$. Thus $s_n \to x_0$. Q.E.D.

Exercises

1. Let $\{c_n\}$ be a sequence of real numbers, and let $\{G_n\}$ be a sequence of distribution functions defined by

$$G_n(x) = \begin{cases} 0 & \text{if } x < c_n \\ 1 & \text{if } x \geq c_n. \end{cases}$$

Suppose $G_n \xrightarrow{c} G$. Show that there is a number L such that $c_n \to L$ as $n \to \infty$.

2. Prove that if a distribution function is symmetric, then 0 is a median.

3. Let X be a random variable. Prove that if $E \mid X \mid < \infty$, then

$$E \mid X - \mu X \mid \leq E \mid X - c \mid$$

for every real c.

4. Prove the remark on the bottom of page 119.

5.3. The Strong Law of Large Numbers

The strong law of large numbers is the name applied to a class of theorems which deal with almost sure convergence of the sequence of arithmetic means of a sequence of random variables. The outstanding theorem of this class is Kolmogorov's strong law of large numbers which states that if the sequence of random variables is independent and identically distributed, then the sequence of arithmetic means converges almost surely if and only if their common expectation exists, in which case the limit random variable is a constant and is equal to this expectation with probability one. This section is devoted to the proof of this theorem.

Lemma 1. If $\{a_n\}$ is a sequence of real numbers, and if $a_n \to a$ as $n \to \infty$, then $(a_1 + \cdots + a_n)/n \to a$ as $n \to \infty$.

Proof: Let $\epsilon > 0$ be arbitrary. Then there exists an n_ϵ such that if $n > n_\epsilon$, then $\mid a_n - a \mid < \epsilon$. Denote $\bar{a}_n = (a_1 + \cdots + a_n)/n$, and note that for $n > n_\epsilon$

$$\bar{a}_n = \frac{a_1 + \cdots + a_{n_\epsilon}}{n} + \frac{n - n_\epsilon}{n} \cdot \frac{a_{n_\epsilon+1} + \cdots + a_n}{n - n_\epsilon},$$

which yields

$$\limsup_{n \to \infty} \bar{a}_n \leq a + \epsilon \qquad \text{and} \qquad \liminf_{n \to \infty} a_n \geq a - \epsilon,$$

from which the conclusion follows.

Lemma 2 (Kronecker's Lemma). If $\sum_{n=1}^{\infty} b_n$ is a convergent series of real numbers, then $(1/n) \sum_{k=1}^{n} k b_k \to 0$ as $n \to \infty$.

Proof: Let $b = \sum_{k=1}^{\infty} b_k$. Then by Lemma 1,

$$\frac{b_1 + (b_1 + b_2) + \cdots + (b_1 + \cdots + b_n)}{n} \to b \qquad \text{as } n \to \infty,$$

or

$$\frac{nb_1 + (n-1)b_2 + \cdots + b_n}{n} = \frac{n+1}{n} \sum_{k=1}^{n} b_k$$

$$-\frac{1}{n} \sum_{k=1}^{n} kb_k \to b \qquad \text{as } n \to \infty,$$

which implies the conclusion.

Lemma 3. If $\{X_n\}$ are independent random variables, and if $(X_1 + \cdots + X_n)/n$ converges almost surely as $n \to \infty$ to some random variable X, then for some constant c, $P[X = c] = 1$.

Proof: For each fixed k, $X_k/n \overset{\text{a.s.}}{\to} 0$. Hence the hypothesis implies that

$$\frac{(X_k + X_{k+1} + \cdots + X_n)}{n} \overset{\text{a.s.}}{\to} X \qquad \text{for every } k.$$

Hence X is measurable with respect to the tail sigma field of $\{X_n\}$, and by Corollary 1 to Theorem 3 in Section 3.3 we obtain the conclusion.

Lemma 4. If X is a random variable, then EX exists if and only if $\sum_{n=1}^{\infty} P[\,|X| \geq n\,] < \infty$.

Proof: This follows from the following two inequalities:

$$\int |X|\, dP = \int \left(\sum_{n=1}^{\infty} |X|\, I_{[n-1 \leq |X| < n]} \right) dP$$

$$\leq \sum_{n=1}^{\infty} nP[n-1 \leq |X| < n] = \sum_{n=0}^{\infty} P[\,|X| \geq n\,],$$

and

$$\int |X|\, dP \geq \sum_{n=0}^{\infty} P[\,|X| \geq n\,] - 1. \qquad \text{Q.E.D.}$$

Theorem 1 (Kolmogorov's Sufficient Conditions for the Strong Law of Large Numbers). If $\{X_n\}$ are independent random variables with finite second moments, and if $\sum_{n=1}^{\infty} \text{Var}\,(X_n)/n^2 < \infty$, then $(1/n)\sum_{k=1}^{n}(X_k - EX_k) \xrightarrow{\text{a.s.}} 0$.

Proof: Since $\sum_{n=1}^{\infty} \text{Var}\,(X_n)/n^2 < \infty$, then by Theorem 2 in Section 5.1, it follows that $\sum_{n=1}^{\infty}(X_n - EX_n)/n$ converges almost surely. Applying Lemma 2 we may conclude that $(1/n)\sum_{k=1}^{n}(X_k - EX_k)$ converges almost surely.

Theorem 2 (Kolmogorov's Strong Law of Large Numbers). Let $\{X_n\}$ be a sequence of independent identically distributed random variables. Then $(1/n)\sum_{k=1}^{n}X_k$ converges almost surely if and only if EX_1 exists, in which case $(1/n)\sum_{k=1}^{n}X_k \xrightarrow{\text{a.s.}} EX_1$.

Proof: Assume $(1/n)\sum_{k=1}^{n}X_k \xrightarrow{\text{a.s.}}$. Define $S_n = \sum_{k=1}^{n}X_k$; then

$$\frac{X_n}{n} = \frac{S_n}{n} - \frac{S_{n-1}}{(n-1)}\frac{(n-1)}{n} \xrightarrow{\text{a.s.}} 0,$$

or $X_n/n \xrightarrow{\text{a.s.}} 0$. Hence $P[\,|\,X_n/n\,| \geq 1 \text{ i.o.}] = 0$, or

$$P[\,|\,X_n\,| \geq n \text{ i.o.}] = 0.$$

By the Borel lemmas (Theorem 2 of Section 3.3), since the events $\{[\,|\,X_n\,| \geq n]\}$ are independent, then $\sum_{n=1}^{\infty}P[\,|\,X_n\,| \geq n] < \infty$. But $P[\,|\,X_n\,| \geq n] = P[\,|\,X_1\,| \geq n]$, since $\{X_n\}$ are identically distributed. Thus, $\sum_{n=1}^{\infty}P[\,|\,X_1\,| \geq n] < \infty$, which by Lemma 4 implies that EX_1 exists.

Conversely, assume that EX_1 exists. If we denote $Y_n = X_n I_{[|X_n| < n]}$, then $P[Y_n \neq X_n] = P[\,|\,X_n\,| \geq n] = P[\,|\,X_1\,| \geq n]$, and thus by Lemma 4, $\sum_{n=1}^{\infty}P[Y_n \neq X_n] < \infty$. By the Borel-Cantelli lemma, then, we have $P[Y_n \neq X_n \text{ i.o.}] = 0$, and hence it is sufficient to prove that $(1/n)\sum_{k=1}^{n}Y_k \to EX_1$. Since $EY_n = E(X_1 I_{[|X_1| < n]}) \to E(X_1)$ as $n \to \infty$, then by Lemma 1, $(EY_1 + \cdots + EY_n)/n \to EX_1$ as $n \to \infty$. Hence it suffices to prove that $(1/n)\sum_{k=1}^{n}(Y_k - EY_k) \xrightarrow{\text{a.s.}} 0$. By Theorem 1 it is sufficient to prove that $\sum_{k=1}^{\infty} \text{Var}\,(Y_k)/k^2 < \infty$. Since $\text{Var}\,(Y_k) = E(Y_k^2) - E^2(Y_k) \leq E(Y_k^2)$, it is sufficient to prove $\sum_{n=1}^{\infty}E(Y_n^2)/n^2 < \infty$.

Now $E(Y_n{}^2) = E(X_1{}^2 I_{[|X_1|<n]})$, so it is sufficient to prove that

$$\sum_{n=1}^{\infty} \frac{E(X_1{}^2 I_{[|X_1|<n]})}{n^2} < \infty.$$

By the monotone convergence theorem it is sufficient to prove that

$$\int \left(\sum_{n=1}^{\infty} \frac{X_1{}^2 I_{[|X_1|<n]}}{n^2} \right) dP < \infty.$$

But, for $m = 1, 2, \cdots,$

$$\int \left(\sum_{n=1}^{\infty} \frac{X_1{}^2 I_{[|X_1|<n]}}{n^2} \right) I_{[m-1\leq|X_1|<m]} \, dP$$

$$= \int \left(\sum_{n=m}^{\infty} \frac{X_1{}^2}{n^2} \right) I_{[m-1\leq|X_1|<m]} \, dP$$

$$\leq \int m^2 \left(\sum_{n=m}^{\infty} n^{-2} \right) I_{[m-1\leq|X_1|<m]} \, dP$$

$$\leq m^2 \frac{1}{(m-1)} P[m-1 \leq |X_1| < m]$$

$$< 2mP[m-1 \leq |X_1| < m].$$

Hence

$$\int \left(\sum_{n=1}^{\infty} \frac{X_1{}^2 I_{[|X_1|<n]}}{n^2} \right) dP$$

$$\leq 2 \sum_{m=1}^{\infty} mP[m-1 \leq |X_1| < m] \leq 2(E|X_1| + 1) < \infty,$$

which concludes the proof.

EXERCISE

1. If $x \in [0, 1]$ and if n is a positive integer, then x can be represented in a unique manner by

$$x = \frac{a_{n,1}(x)}{n} + \frac{a_{n,2}(x)}{n^2} + \cdots,$$

where $a_{n,j}(x) = 0$ or 1 or $\cdots$ or $n - 1$, and where for fixed n and fixed x it is not true that all but a finite number of $\{a_{n,j}(x), j = 1, 2, \cdots\}$ are $n - 1$. A number $x \in [0, 1]$ is said to be *normal* if for every n and every $j \in \{0, 1, \cdots, n - 1\}$,

$$\frac{\text{the number of } j \text{ among } \{a_{n,1}(x), \cdots, a_{n,N}(x)\}}{N} \to \frac{1}{n}$$

as $N \to \infty$. Prove Borel's celebrated theorem which states: the set of normal numbers in $[0, 1]$ has Lebesgue measure 1 (that is, almost all numbers are normal).

5.4. The Glivenko-Cantelli Theorem

Often in the statistical literature one reads the statement, "a random sample describes the population." What does this mean in crisp measure-theoretic terms? What it means is the following: Let $(\Omega, \mathcal{C}, P)$ be a probability space, X a random variable, and F its distribution function. Now let $\{X_n\}$ be a sequence of independent identically distributed random variables with common distribution function F. We say that $\{X_n\}$ is *a sequence of independent observations on X*. The n random variables $X_1, \cdots, X_n$ are referred to as a sample (or random sample) of size n on the "population" which here is indistinguishable from F. The problem, then, is to reconstruct the so-called population or distribution function F from the observable random variables $\{X_n\}$. The theorem which says that this can be done, *and in a uniform manner*, is the Glivenko-Cantelli theorem which is given in this section. This theorem is the meaning of the statement, "a random sample describes a population."

For this section $\{X_n\}$ will be an arbitrary sequence of independent identically distributed random variables with arbitrary common distribution function F.

Definition: The nth empirical distribution function $\widehat{F}_n(x)$ of $F(x)$ is defined to be $1/n$ times the number of random variables among $\{X_1, \cdots, X_n\}$ which are $\leq x$. Equivalently, $\widehat{F}_n(x)$ is the proportion among the first n random variables which are $\leq x$.

Remark: A precise way of writing $\widehat{F}_n(x)$ is

$$\widehat{F}_n(x) = \frac{1}{n} \sum_{k=1}^{n} I_{[X_k \leq x]}.$$

Lemma 1. For every real x, $\widehat{F}_n(x) \to F(x)$ almost surely and

$$\widehat{F}_n(x - 0) \to F(x - 0)$$

almost surely as $n \to \infty$.

Proof: Since $F(x) = P[X_1 \leq x]$, $F(x - 0) = P[X_1 < x]$, $\widehat{F}_n(x) = (1/n) \sum_{k=1}^{n} I_{[X_k \leq x]}$, and $\widehat{F}_n(x - 0) = (1/n) \sum_{k=1}^{n} I_{[X_k < x]}$, we apply the strong law of large numbers to the two sequences of independent identically distributed random variables $\{I_{[X_n \leq x]}\}$ and $\{I_{[X_n < x]}\}$ and obtain the conclusion of the lemma.

It should be noted that the conclusions of Lemma 1 hold for arbitrary x. There are now two questions which arise. *Question 1*: Does $P[\widehat{F}_n(x) \to F(x)$ for all real $x] = 1$? Note that here we are taking the probability of an uncountable intersection of events, and we have no guarantee of its being an event. Even so, suppose it is an event and the answer is yes. Then we can ask an even stronger question. *Question 2*: Does $P[\widehat{F}_n(x) \to F(x)$ uniformly in $x] = 1$? The Glivenko-Cantelli theorem, sometimes referred to as the *fundamental theorem of statistics*, says that there is an affirmative answer to question number 2.

Theorem 1 (Glivenko-Cantelli Theorem). Let $\{X_n\}$ be a sequence of independent, identically distributed random variables with arbitrary common distribution function F. Then

$$P[\sup_{-\infty < x < \infty} |\widehat{F}_n(x) - F(x)| \to 0] = 1.$$

Proof: Let r be any positive integer ≥ 2. For $k = 1, 2, \cdots, r - 1$, define

$$x_{r,k} = \min \{x \mid k/r \leq F(x)\},$$

and define $x_{r,0} = -\infty$ and $x_{r,r} = +\infty$. We need only consider in the sequel those intervals $[x_{r,k}, x_{r,k+1})$ which are *nonempty*. Now, if

$$x \in [x_{r,k}, x_{r,k+1}),$$

then

$$\hat{F}_n(x) - F(x) \leq \hat{F}_n(x_{r,k+1} - 0) - F(x_{r,k})$$

$$= (\hat{F}_n(x_{r,k+1} - 0) - F(x_{r,k+1} - 0))$$

$$+ (F(x_{r,k+1} - 0) - F(x_{r,k}))$$

$$\leq \hat{F}_n(x_{r,k+1} - 0) - F(x_{r,k+1} - 0) + r^{-1}.$$

Also

$$\hat{F}_n(x) - F(x) \geq \hat{F}_n(x_{r,k}) - F(x_{r,k+1} - 0)$$

$$= \hat{F}_n(x_{r,k}) - F(x_{r,k})$$

$$- (F(x_{r,k+1} - 0) - F(x_{r,k}))$$

$$\geq \hat{F}_n(x_{r,k}) - F(x_{r,k}) - r^{-1}.$$

It should be noticed that these two inequalities hold for every $\omega \in \Omega$. Thus for every real x,

$$| \hat{F}_n(x) - F(x) | \leq \max_{\substack{1 \leq k \leq r-1 \\ 1 \leq j \leq r-1}} \{ | \hat{F}_n(x_{r,k}) - F(x_{r,k}) |, | \hat{F}_n(x_{r,j} - 0)$$

$$- F(x_{r,j} - 0) | \} + r^{-1}.$$

The inequality holds if we take supremum of both sides over all x. We now take lim sup of both sides as $n \to \infty$ and use Lemma 1 to obtain

$$\limsup_{n \to \infty} \sup_{-\infty < x < +\infty} | \hat{F}_n(x) - F(x) | \leq r^{-1} \text{ a.s.},$$

and by the arbitrariness and countability of the values of r we obtain the conclusion.

It should be noticed that the above proof was accomplished without ever verifying whether $\sup_{-\infty < x < \infty} | \hat{F}_n(x) - F(x) |$ is a random variable or not.

EXERCISES

1. Prove that $\sup_{-\infty < x < \infty} | \hat{F}_n(x) - F(x) |$ as defined above is a random variable.

2. Let $\{X_n\}$ be a sequence of independent, identically distributed random variables with common distribution function F. Let $0 < p < 1$ and x be such that $F(x - \epsilon) < p < F(x + \epsilon)$ for all $\epsilon > 0$. Let $X_{n,1}, \cdots, X_{n,n}$ be defined by $X_{n,j}$ as the jth smallest of $\{X_1, \cdots, X_n\}$; that is, $X_{n,1}(\omega) = \min\{X_1(\omega), \cdots, X_n(\omega)\}$, $X_{n,2}(\omega)$ is the next to smallest of $\{X_1(\omega), \cdots, X_n(\omega)\}$, etc., for every ω.

(a) Prove that $X_{n,1}, \cdots, X_{n,n}$ are random variables, and

(b) If $[u]$ denotes the largest integer $\leq u$, prove that $X_{n,[np]} \to x$ a.s.

5.5. Inequalities for the Law of the Iterated Logarithm

The inequalities proved in this section are used to prove the so-called law of the iterated logarithm in Section 5.6. Two theorems due to A. N. Kolmogorov are proved. In Theorem 1 two inequalities are proved; the first one is used directly in the proof of the law of the iterated logarithm, and both are used in the proof of Theorem 2, which itself is used for the same purpose.

We first take note of a very easy inequality.

Remark: If $t \geq 0$, then

(1) $$\exp[t(1 - t)] \leq 1 + t \leq e^t.$$

Proof: The second inequality is trivial. Since the derivative of $\exp[t(1 - t)]$ is negative for $t > \frac{1}{2}$, we need only prove the first inequality for $t \in (0, \frac{1}{2})$. At $t = 0$, both $\exp[t(1 - t)]$ and $1 + t$ are equal and their derivatives are equal. However, for $t \in (0, \frac{1}{2})$ the second derivative of $\exp[t(1 - t)]$ is negative, it being $\exp[t(1 - t)]((1 - 2t)^2 - 2)$, while the second derivative of $1 + t$ is 0. Q.E.D.

In Theorems 1 and 2 in this section the following notation will be used. We let $X_1, \cdots, X_n$ denote n independent, bounded random variables satisfying $EX_k = 0, 1 \leq k \leq n$. We denote $S_n = X_1 + \cdots + X_n$, $s_n^2 = \text{Var } S_n$, $s_n = \sqrt{s_n^2}$, and

$$c_n = \max\{\text{ess sup } |X_k|/s_n \mid 1 \leq k \leq n\}.$$

Theorem 1. Using the above assumptions and notation, we have:

(a) if $0 < \epsilon < 1/c_n$, then

$$P\left[\frac{S_n}{s_n} > \epsilon\right] < \exp\left\{-\frac{\epsilon^2}{2}\left(1 - \frac{\epsilon c_n}{2}\right)\right\},$$

and

(b) if $\epsilon \geq 1/c_n$, then

$$P\left[\frac{S_n}{s_n} > \epsilon\right] < \exp\left\{\frac{-\epsilon}{4c_n}\right\}.$$

Proof: Let X be a random variable and $c > 0$ be a number such that $EX = 0$ and $P[\,|X| \leq c] = 1$. Then, if $\sigma^2 = EX^2$, we have for $t \geq 0$,

$$E[\exp(tX)] = 1 + \frac{t^2EX^2}{2!} + \frac{t^3EX^3}{3!} + \cdots$$

$$\leq 1 + \frac{t^2\sigma^2}{2}\left(1 + \frac{tc}{3} + \frac{t^2c^2}{3\cdot4} + \cdots\right)$$

$$= 1 + \frac{t^2\sigma^2}{2}\left(1 + \frac{tc}{3}\left(1 + \frac{tc}{4} + \cdots\right)\right).$$

If we require $0 < tc \leq 1$, then

$$1 + \frac{tc}{4} + \frac{t^2c^2}{4\cdot5} + \cdots < \sum_{n=0}^{\infty} 4^{-n} = \tfrac{4}{3}.$$

Hence for $0 < tc \leq 1$,

$$E[\exp(tX)] < 1 + \frac{t^2\sigma^2}{2}\left(1 + \frac{tc}{2}\right),$$

and because of the second inequality in (1) we obtain

(2) $$E[\exp(tX)] < \exp\left\{\frac{t^2\sigma^2}{2}\left(1 + \frac{tc}{2}\right)\right\}.$$

Also, for $0 < tc \leq 1$, and since $EX^n \geq -c^{n-2} EX^2$ for $n \geq 3$, we have

$$E[\exp(tX)] \geq 1 + \frac{t^2\sigma^2}{2}\left(1 - \frac{tc}{3} - \frac{t^2c^2}{3\cdot4} - \cdots\right)$$

from which, by the steps leading up to (2), we obtain

(3) $$E[\exp(tX)] > 1 + \frac{t^2\sigma^2}{2}\left(1 - \frac{tc}{2}\right).$$

By the first inequality in (1), we have

(4) $\quad 1 + \dfrac{t^2\sigma^2}{2}\left(1 - \dfrac{tc}{2}\right) > \exp\left\{\dfrac{t^2\sigma^2}{2}\left(1 - \dfrac{tc}{2}\right)\left(1 - \dfrac{t^2\sigma^2}{2}\left(1 - \dfrac{tc}{2}\right)\right)\right\}.$

But since $0 \le \sigma^2 \le c^2$, then

(5) $\quad \left(1 - \dfrac{tc}{2}\right)\left(1 - \dfrac{t^2\sigma^2}{2}\left(1 - \dfrac{tc}{2}\right)\right) > 1 - \dfrac{tc}{2} - \dfrac{t^2\sigma^2}{2} \ge 1 - tc.$

From (3), (4), and (5) we obtain

(6) $\quad E[\exp\,(tX)] > \exp\left\{\dfrac{t^2\sigma^2}{2}\,(1 - tc)\right\} \qquad$ for $0 < tc \le 1.$

We now prove (a). Independence of $X_1, \cdots, X_n$ implies

$$E \exp\left\{\dfrac{tS_n}{s_n}\right\} = \prod_{k=1}^{n} E \exp\left\{\dfrac{tX_k}{s_n}\right\}.$$

By (6) and (2) we have

$$\exp\left\{\dfrac{(\operatorname{Var} X_k)t^2}{2s_n^2}\,(1 - tc_n)\right\} < E \exp\left\{\dfrac{tX_k}{s_n}\right\}$$

$$< \exp\left\{\dfrac{t^2 \operatorname{Var} X_k}{2s_n^2}\left(1 + \dfrac{tc_n}{2}\right)\right\}.$$

Taking the product of each term in this last inequality from $k = 1$ to n, and remembering that $\operatorname{Var} S_n = \sum_{k=1}^{n} \operatorname{Var} X_k = s_n^2$, we obtain

(7) $\quad \exp\left\{\dfrac{t^2}{2}\,(1 - tc_n)\right\} < E \exp\left\{\dfrac{tS_n}{s_n}\right\} < \exp\left\{\dfrac{t^2}{2}\left(1 + \dfrac{tc_n}{2}\right)\right\}.$

One easily obtains

$$E \exp\left\{\dfrac{tS_n}{s_n}\right\} \ge E\left(I_{[S_n/s_n>\epsilon]} \exp\left\{\dfrac{tS_n}{s_n}\right\}\right)$$

$$\ge \exp\,(t\epsilon)P\left[\dfrac{S_n}{s_n} > \epsilon\right],$$

which, together with the second inequality in (7) yields

(8) $\quad P\left[\dfrac{S_n}{s_n} > \epsilon\right] \le \exp\,(-t\epsilon)E \exp\left\{\dfrac{tS_n}{s_n}\right\} < \exp\left\{-t\epsilon + \dfrac{t^2}{2}\left(1 + \dfrac{tc_n}{2}\right)\right\}.$

If $0 < \epsilon \leq 1/c_n$, we replace t by ϵ and obtain

$$P\left[\frac{S_n}{s_n} > \epsilon\right] < \exp\left\{-\epsilon^2 + \frac{\epsilon^2}{2} + \frac{\epsilon^3 c_n}{4}\right\}$$

$$= \exp\left\{-\frac{\epsilon_2}{2}\left(1 - \frac{\epsilon c_n}{2}\right)\right\},$$

which proves inequality (a) in the statement of the theorem. If $\epsilon \geq 1/c_n$, replace t by $1/c_n$ in (8) (so that $tc_n \leq 1$), and thus

$$P\left[\frac{S_n}{s_n} > \epsilon\right] < \exp\left\{-\frac{\epsilon}{c_n} + \frac{1}{2c_n{}^2} + \frac{1}{4c_n{}^2}\right\}.$$

Since

$$-\frac{\epsilon}{c_n} + \frac{3}{4c_n{}^2} \leq -\frac{\epsilon}{c_n} + \frac{3\epsilon}{4c_n} = -\frac{\epsilon}{4c_n},$$

then

$$P\left[\frac{S_n}{s_n} > \epsilon\right] < \exp\left\{-\frac{\epsilon}{4c_n}\right\},$$

which verifies inequality (b) in the statement of the theorem.

Theorem 2. If $\gamma > 0$ is arbitrarily selected, if $c_n = c_n(\gamma)$ is sufficiently small, and if $\epsilon = \epsilon(\gamma)$ is sufficiently large, then

$$P\left[\frac{S_n}{s_n} > \epsilon\right] > \exp\left\{-\frac{\epsilon^2(1 + \gamma)}{2}\right\}.$$

Proof: Let $\{\alpha, \beta\} \subset (0, 1)$. Then by the first inequality in (7), if c_n is sufficiently small, namely, if c_n satisfies $0 < c_n < \alpha/t$, then

(9) $$E \exp\left\{\frac{tS_n}{s_n}\right\} > \exp\left\{\frac{t^2(1 - \alpha)}{2}\right\}.$$

Let

$$q(x) = P\left[\frac{S_n}{s_n} > x\right].$$

Integrating by parts we get

$$E \exp \left(\frac{tS_n}{s_n} \right) = - \int_{-\infty}^{\infty} \exp (tx) \, dq(x)$$

$$= - \exp (tx) q(x) \Big|_{-\infty}^{+\infty} + t \int_{-\infty}^{\infty} \exp (tx) q(x) \, dx.$$

Since S_n is bounded, then $q(x) = 0$ for x sufficiently large, and thus

$$E \exp \left(\frac{tS_n}{s_n} \right) = t \int_{-\infty}^{\infty} \exp (tx) q(x) \, dx.$$

Let $I_1 = (-\infty, 0]$, $I_2 = (0, t(1 - \beta)]$, $I_3 = (t(1 - \beta), t(1 + \beta)]$, $I_4 = (t(1 + \beta), 8t]$, and $I_5 = (8t, \infty)$. Let us denote

$$J_i = t \int_{I_i} \exp (tx) q(x) \, dx, \qquad i = 1, 2, 3, 4, 5.$$

We first observe that

$$J_1 = t \int_{-\infty}^{0} \exp (tx) q(x) \, dx < t \int_{-\infty}^{0} \exp (tx) \, dx = 1.$$

We next evaluate J_5. Let c_n be small enough so that $8tc_n < 1$. Hence $1/4c_n > 2t$ or $-1/4c_n < -2t$. Hence for $x \geq 1/c_n$ we obtain from (b) in Theorem 1 that

$$q(x) = P \left[\frac{S_n}{s_n} > x \right] < \exp (-x/4c_n) < \exp (-2tx).$$

But for $0 < x < 1/c_n$ (or $0 < xc_n < 1$) we have by (a) in Theorem 1 that

$$q(x) = P \left[\frac{S_n}{s_n} > x \right] < \exp \left\{ -\frac{x^2}{2} \left(1 - \frac{xc_n}{2} \right) \right\} < \exp (-x^2/4).$$

Hence if $8t < x < 1/c_n$ (since $x \in I_5$), then since $0 < \exp (-x/4) < 1$, we have

$$\exp (-x^2/4) = [\exp (-x/4)]^x < [\exp (-x/4)]^{8t} = \exp (-2xt).$$

Thus for all $x \in I_5$, $q(x) < \exp (-2xt)$ for c_n sufficiently small, and

$$J_5 = t \int_{8t}^{\infty} \exp (tx) q(x) \, dx < t \int_{8t}^{\infty} \exp (-tx) \, dx < 1.$$

Hence

(10) $J_1 + J_5 < 2.$

We next consider J_2 and J_4. If c_n is sufficiently small, we have $0 < x < 1/c_n$ over I_2 and I_4. Hence, by (a) in Theorem 1,

$$\exp (tx) q(x) \; < \; \exp \left\{ tx - \left(\frac{x^2}{2}\right)\left(1 - \frac{xc_n}{2}\right) \right\}.$$

Over $I_2 \cup I_4$, $x \leq 8t$, so

$$\exp (tx) q(x) \; < \; \exp \left\{ tx - \left(\frac{x^2}{2}\right)(1 - 4tc_n) \right\} \; = \; \exp g(x),$$

where $g(x)$ is defined by the last equality. Now $g(x)$ is a quadratic in x and achieves its maximum value when $t - x(1 - 4tc_n) = 0$ or at $x_0 = t/(1 - 4tc_n)$. For c_n sufficiently small, x_0 will lie in $I_3 = (t(1 - \beta), t(1 + \beta)]$, which separates I_2 and I_4. Hence for c_n sufficiently small, if $x \in I_2$, then

$$g(x) \leq g(t(1 - \beta))$$

$$= t^2(1 - \beta) - \frac{t^2(1 - \beta)^2}{2} + 2t^3(1 - \beta)^2 c_n$$

$$= \left(\frac{t^2}{2}\right)(1 - \beta^2 + 4tc_n(1 - \beta)^2).$$

For c_n sufficiently small (β being preselected in $(0, 1)$), we have

$$4tc_n(1 - \beta)^2 < \beta^2/2.$$

Hence $g(x) < (t^2/2)(1 - \beta^2/2)$ for $x \in I_2$, and

$$J_2 = t \int_0^{t(1-\beta)} \exp (tx) q(x) \, dx < t \int_0^{t(1-\beta)} \exp \{g(x)\} \, dx$$

$$< t^2(1 - \beta) \exp \left\{ \left(\frac{t^2}{2}\right)\left(1 - \frac{\beta^2}{2}\right) \right\}$$

$$< t^2 \exp \left\{ \left(\frac{t^2}{2}\right)\left(1 - \frac{\beta^2}{2}\right) \right\}.$$

Also, for $x \in I_4$,

$$g(x) \le t^2(1 + \beta) - \frac{t^2(1 + \beta)^2}{2} + 2t^3c_n(1 + \beta)^2$$

$$= \frac{t^2}{2}(1 - \beta^2 + 4tc_n(1 + \beta)^2).$$

For c_n sufficiently small, $4tc_n(1 + \beta)^2 < \beta^2/2$. Hence

$$g(x) \le (t^2/2)(1 - \beta^2/2),$$

and we obtain

$$J_4 = t \int_{t(1+\beta)}^{8t} \exp(tx)q(x)\,dx$$

$$< t \int_{t(1+\beta)}^{8t} \exp[g(x)]\,dx < 8t^2 \exp\left\{\frac{t^2}{2}\left(1 - \frac{\beta^2}{2}\right)\right\}.$$

Whatever be $\beta \in (0, 1)$, let us select α by $\alpha = \beta^2/4$. Also, let

$$t = \epsilon/(1 - \beta).$$

Then

$$J_2 + J_4 < 9t^2 \exp\left\{\frac{t^2}{2}\left(1 - \frac{\beta^2}{2}\right)\right\}$$

$$= 9t^2 \exp\left\{\frac{-\beta^2 t^2}{8}\right\} \exp\left\{\frac{t^2}{2}\left(1 - \frac{\beta^2}{4}\right)\right\}.$$

Hence for ϵ sufficiently large (hence t *is* large) we have, by (9),

$$J_2 + J_4 < \tfrac{1}{4}E \exp\{tS_n/s_n\}.$$

We make use now of the fact that $ES_n = 0$. Thus as ϵ gets larger but β remains fixed, t gets larger and $E \exp(tS_n/s_n)$ can become large since $P[S_n > 0] > 0$. Hence $2 < \tfrac{1}{4}E \exp\{tS_n/s_n\}$ for sufficiently large ϵ (or t), and thus by (10),

(11) $$J_1 + J_5 < \tfrac{1}{4}E \exp\left\{\frac{tS_n}{s_n}\right\}.$$

Since $J_1 + J_2 + J_3 + J_4 + J_5 = E \exp \{tS_n/s_n\}$, it follows that

(12)
$$J_3 > \tfrac{1}{2}E \exp \left| \frac{tS_n}{s_n} \right|.$$

But

$$J_3 = t \int_{t(1-\beta)}^{t(1+\beta)} \exp (tx) q(x) \, dx$$

$$\leq 2t^2 \beta \exp [t^2(1 + \beta)] q(t(1 - \beta)),$$

and since $t = \epsilon/(1 - \beta)$, we obtain

(13)
$$J_3 \leq 2t^2 \beta \exp [t^2(1 + \beta)] q(\epsilon).$$

By (13), (12) and (9) we obtain

$$2t^2 \beta \exp [t^2(1 + \beta)] q(\epsilon) > \tfrac{1}{2} \exp \left| \frac{t^2(1 - \alpha)}{2} \right|.$$

Thus

$$q(\epsilon) > \left(\frac{1}{4t^2\beta} \right) \exp \left| \frac{t^2}{2} - \frac{\alpha t^2}{2} - t^2 - \beta t^2 \right|$$

$$= (4t^2 \beta)^{-1} \exp \left| \frac{t^2\alpha}{2} \right| \exp \left\{ -\frac{t^2}{2} - \alpha t^2 - \beta t^2 \right\}$$

$$= (4t^2 \beta)^{-1} \exp \left| \frac{t^2\alpha}{2} \right| \exp \tfrac{1}{2} \{ -t^2(1 + 2\alpha + 2\beta) \}.$$

Now $(4t^2\beta)^{-1} \exp \{t^2\alpha/2\}$ tends to ∞ as ϵ (or t) becomes sufficiently large, and hence for sufficiently large ϵ,

$$q(\epsilon) > \exp \tfrac{1}{2} \{ -t^2(1 + 2\alpha + 2\beta) \}$$

$$= \exp \left\{ -\frac{\epsilon^2}{2} \cdot \frac{(1 + 2\alpha + 2\beta)}{(1 - \beta)^2} \right\}$$

since $t = \epsilon(1 - \beta)$. But given $\gamma > 0$, we can select $\beta > 0$ sufficiently small such that (recalling that $\alpha = \beta^2/4$)

$$\frac{(1 + 2\beta + \tfrac{1}{2}\beta^2)}{(1 - \beta)^2} \leq 1 + \gamma.$$

Hence $q(\epsilon) > \exp \{ -\epsilon^2(1 + \gamma)/2 \}$ for c_n sufficiently small and ϵ sufficiently large, which proves the theorem.

5.6. The Law of the Iterated Logarithm

In this section we prove one of the sharpest strong limit theorems, known as "the law of the iterated logarithm." We first prove several lemmas.

Lemma 1. If X is a random variable with finite second moment, then $| \mu X - EX | \leq \sqrt{2 \operatorname{Var} X}$ for every median μX.

Proof: By Chebishev's inequality, for every $\delta > 0$,

$$P[\, | X - EX | \leq \sqrt{(2 + \delta) \operatorname{Var} X}\,] \geq 1 - (2 + \delta)^{-1} > \tfrac{1}{2}.$$

Hence

$$P[X \geq EX - \sqrt{(2 + \delta) \operatorname{Var} X}\,] > \tfrac{1}{2},$$

and

$$P[X \leq EX + \sqrt{(2 + \delta) \operatorname{Var} X}\,] > \tfrac{1}{2}.$$

By the definition of median it follows that every median μX of X satisfies $EX - \sqrt{(2 + \delta) \operatorname{Var} X} \leq \mu X \leq EX + \sqrt{(2 + \delta) \operatorname{Var} X}$. The conclusion follows because of the arbitrariness of $\delta > 0$.

Lemma 2 (P. Lévy Inequalities). If $X_1, \cdots, X_n$ are n independent random variables, if $S_k = X_1 + \cdots + X_k$, and if ϵ is any constant, then

(1) $$P[\, \max_{1 \leq k \leq n} (S_k - \mu(S_k - S_n)) \geq \epsilon] \leq 2P[S_n \geq \epsilon],$$

and

(2) $$P[\, \max_{1 \leq k \leq n} | S_k - \mu(S_k - S_n) | \geq \epsilon] \leq 2P[\, | S_n | \geq \epsilon].$$

Proof: We first prove (1). Let us denote

$$S_0 = 0, \quad S_k^* = \max_{1 \leq j \leq k} (S_j - \mu(S_j - S_n)), \qquad 1 \leq k \leq n,$$

$$A_k = [S_{k-1}^* < \epsilon][S_k - \mu(S_k - S_n) \geq \epsilon],$$

and

$$B_k = [S_n - S_k - \mu(S_n - S_k) \geq 0].$$

We may take $\mu(S_n - S_k) = -\mu(S_k - S_n)$. It is clear that $A_1, \cdots, A_n$ are disjoint and that

$$[S_n^* \geq \epsilon] = \bigcup_{k=1}^{n} A_k.$$

However, for every k,

$$B_k A_k \subset [S_n \geq S_k - \mu(S_k - S_n)][S_k - \mu(S_k - S_n) \geq \epsilon]$$

$$\subset [S_n \geq \epsilon].$$

Thus $[S_n \geq \epsilon] \supset \bigcup_{k=1}^{n} A_k B_k$. Since $\{A_1, \cdots, A_n\}$ are disjoint, so are $\{A_1 B_1, \cdots, A_n B_n\}$. Further, by the definition of median, $P(B_k) \geq \frac{1}{2}$. Also, for every k, A_k and B_k are independent. Hence

$$P[S_n \geq \epsilon] \geq P(\bigcup_{k=1}^{n} A_k B_k) = \sum_{k=1}^{n} P(A_k B_k)$$

$$= \sum_{k=1}^{n} P(A_k) P(B_k) \geq \frac{1}{2} \sum_{k=1}^{n} P(A_k)$$

$$= \frac{1}{2} P(\bigcup_{k=1}^{n} A_k) = \frac{1}{2} P[S_n^* \geq \epsilon],$$

which proves (1). In a similar manner one can prove that

$$P(S_n \leq -\epsilon) \geq \frac{1}{2} P[S_n^{**} \leq -\epsilon],$$

where $S_n^{**} = \min_{1 \leq j \leq n}(S_j - \mu(S_j - S_n))$, which combined with (1) yields (2). Q.E.D.

Lemma 3. If $X_1, \cdots, X_n$ are independent random variables with finite second moments, if $EX_k = 0$ for $1 \leq k \leq n$, and if $S_k = X_1 + \cdots + X_k$, then

$$P[\max_{1 \leq k \leq n} S_k \geq \epsilon] \leq 2P[S_n \geq \epsilon - \sqrt{2 \operatorname{Var} S_n}].$$

Proof: Since $E(S_k - S_n) = 0$, then by Lemma 1, $|\mu(S_k - S_n)| \leq \sqrt{2 \operatorname{Var} (S_k - S_n)} \leq \sqrt{2 \operatorname{Var} S_n}$, or $-\mu(S_k - S_n) \geq -\sqrt{2 \operatorname{Var} S_n}$.

By (1) in Lemma 2,

$$P[\max_{1\leq k\leq n} S_k \geq \epsilon] = P(\bigcup_{k=1}^{n} [S_k \geq \epsilon])$$

$$\leq P(\bigcup_{k=1}^{n} [S_k - \mu(S_k - S_n) \geq \epsilon - \sqrt{2\,\mathrm{Var}\,S_n}])$$

$$= P[\max_{1\leq k\leq n} (S_k - \mu(S_k - S_n)) \geq \epsilon - \sqrt{2\,\mathrm{Var}\,S_n}]$$

$$\leq 2P[S_n \geq \epsilon - \sqrt{2\,\mathrm{Var}\,S_n}],$$

which proves the lemma.

The following two lemmas are elementary properties of sequences. Their proofs are included for the sake of completeness.

Lemma 4. Let $\{b_n\}$ be a nondecreasing sequence of positive numbers such that $b_n \to \infty$ and $b_{n+1}/b_n \to 1$ as $n \to \infty$. For any $c > 1$ there is an (eventually) increasing sequence of positive integers $\{n_k\}$ such that $b_{n_k} \sim c^k$ (that is, $b_{n_k}/c^k \to 1$ as $k \to \infty$).

Proof: First we note that there exists an integer k_0 such that every interval of the form $[c^k, c^{k+1})$ contains at least one b_n for all $k > k_0$; otherwise, $\limsup_{n\to\infty}(b_{n+1}/b_n) \geq c > 1$, which contradicts the hypothesis that $b_{n+1}/b_n \to 1$ as $n \to \infty$. Let $n_1 = n_2 = \cdots = n_{k_0} = 1$. For $j > k_0$ define n_j to be the smallest integer such that $b_{n_j} \geq c^j$. Then $n_j < n_{j+1}$ for all $j > k_0$, so that $\{n_j\}$ *is* eventually increasing. For $j > k_0$, $b_{n_j-1}/b_{n_j} < c^j/b_{n_j} \leq 1$. Since $b_{n_j-1}/b_{n_j} \to 1$, it follows that $b_{n_j} \sim c^j$. Q.E.D.

Lemma 5. If $0 < b_n$ and $b_n \to \infty$ as $n \to \infty$, and if $b_n/b_{n-1} \sim c \geq 1$, then $(\log b_n)/(\log b_{n-1}) \to 1$ as $n \to \infty$.

Proof: We observe that

$$\frac{\log b_n}{\log b_{n-1}} = \left(\frac{\log (b_n/b_{n-1}) + \log b_{n-1}}{\log b_{n-1}}\right).$$

Since $\log (b_n/b_{n-1}) \to \log c$ as $n \to \infty$ and since $\log b_n \to \infty$ as $n \to \infty$, we obtain the conclusion.

Theorem 1 (Kolmogorov's "Law of the Iterated Logarithm").
Let $\{X_n\}$ be a sequence of independent random variables such that
$d_n \doteq \text{ess sup} \,| X_n | < \infty$ and $EX_n = 0$ for all n, and such that

$$s_n^2 \doteq \text{Var} \, S_n \to \infty \quad \text{and} \quad d_n = o(\sqrt{s_n^2/\log\log s_n^2}) \qquad \text{as} \quad n \to \infty,$$

where $S_n = X_1 + \cdots + X_n$. Then

$$P\left[\limsup_{n\to\infty} \frac{S_n}{\sqrt{2s_n^2 \log\log s_n^2}} = 1\right] = 1.$$

Proof: We first observe that

$$\text{Var} \, X_n = EX_n^2 \le d_n^2 = o\left(\frac{s_n^2}{\log\log s_n^2}\right).$$

Hence

$$\frac{s_{n+1}^2}{s_n^2} = 1 + \frac{(\text{Var} \, X_{n+1})}{s_n^2}$$

$$= 1 + \frac{s_{n+1}^2}{s_n^2}\cdot\frac{o(1)}{\log\log s_{n+1}^2},$$

or

$$\frac{s_{n+1}^2}{s_n^2}\left(1 - \frac{o(1)}{\log\log s_{n+1}^2}\right) = 1.$$

Since $\log\log s_n^2 \to \infty$ as $n \to \infty$, this last equation yields:

$$\frac{s_{n+1}^2}{s_n^2} \to 1 \qquad \text{as} \quad n \to \infty.$$

Let ϵ, δ be arbitrarily selected numbers satisfying $1 > \epsilon > \delta > 0$. Let us denote $t_n = (2\log\log s_n^2)^{1/2}$. The theorem will be proved when we prove the following two assertions:

(3) $$P[S_n > (1 + \delta)s_n t_n \text{ i.o.}] = 0,$$

and

(4) $$P[S_n > (1 - \epsilon)s_n t_n \text{ i.o.}] = 1.$$

We first prove (3). Let c satisfy $1 + \delta > c > 1$ and remain fixed. Since we have just shown that $\{s_n\}$ satisfies the hypotheses of Lemma 4, it follows from Lemma 4 that there is an eventually increasing sequence of positive integers $\{n_k\}$ such that $s_{n_k} \sim c^k$. Thus by Lemma 5,

$$(5) \qquad (1 + \delta)s_{n_{k-1}}t_{n_{k-1}} \sim \frac{1 + \delta}{c} s_{n_k}t_{n_k}.$$

Let us denote

$$S^*_{n_k} = \max \{S_j \mid 1 \le j \le n_k\}, \quad \text{and} \quad k(n) = \max \{k \mid n_k \le n\},$$

or

$$n_{k(n)} = \max \{n_k \mid n_k \le n\}.$$

Then, since $\{s_n t_n\}$ is increasing in n,

$$[S_n > (1 + \delta)s_n t_n \text{ i.o.}] \subset [S_n > (1 + \delta)s_{n_{k(n)}}t_{n_{k(n)}} \text{ i.o.}]$$

$$\subset [S^*_{n_{k+1}} > (1 + \delta)s_{n_k}t_{n_k} \text{ i.o.}].$$

Since $1 + \delta > c > 1$, we may select $\zeta > 0$ such that $(1 + \delta)/c > 1 + \zeta$. Because of (5), we have

$$[S^*_{n_{k+1}} > (1 + \delta)s_{n_k}t_{n_k} \text{ i.o.}] \subset [S^*_{n_k} \ge (1 + \zeta)s_{n_k}t_{n_k} \text{ i.o.}].$$

We can prove (3) by proving

$$(6) \qquad P[S^*_{n_k} \ge (1 + \zeta)s_{n_k}t_{n_k} \text{ i.o.}] = 0.$$

By the Borel-Cantelli lemma we need only prove

$$(7) \qquad \sum_{k=1}^{\infty} P[S^*_{n_k} \ge (1 + \zeta)s_{n_k}t_{n_k}] < \infty.$$

By Lemma 3 we have, for all large k,

$$P[S^*_{n_k} \ge (1 + \zeta)s_{n_k}t_{n_k}]$$

$$\le 2P[S_{n_k} \ge (1 + \zeta)s_{n_k}t_{n_k} - \sqrt{2}\, s_{n_k}]$$

$$= 2P\left[S_{n_k} \ge s_{n_k}t_{n_k}\left(1 + \zeta - \frac{\sqrt{2}}{t_{n_k}}\right)\right]$$

$$\le 2P[S_{n_k} \ge s_{n_k}t_{n_k}(1 + \tfrac{1}{2}\zeta)].$$

We now prepare ourselves to apply Theorem 1(a) of Section 5.5. Let us denote $\epsilon_k = t_{n_k}(1 + \zeta/2)$ and

$$c_{n_k} = \max_{1 \le j \le n_k} \text{ess sup} \frac{|X_j|}{s_{n_k}} = o\left(\frac{1}{\sqrt{\log \log s^2_{n_k}}}\right).$$

Since $\epsilon_k c_{n_k} = t_{n_k}(1 + \zeta/2)o(1/t_{n_k}) \to 0$ as $k \to \infty$, it follows that for all large k, $o < \epsilon_k c_{n_k} < 1$. Hence by Theorem 1(a) in Section 5.5, we have

$$P\left[S_{n_k} \geq s_{n_k} t_{n_k}\left(1 + \frac{\varsigma}{2}\right)\right]$$

$$< \exp\left\{-\frac{t^2_{n_k}(1 + \varsigma/2)^2}{2}\left(1 - \frac{\epsilon_k c_{n_k}}{2}\right)\right\}$$

$$= \exp\left\{-\left(1 + \frac{\varsigma}{2}\right)^2\left(1 - \frac{\epsilon_k c_{n_k}}{2}\right)\log\log s^2_{n_k}\right\}.$$

Since, as shown above, $\epsilon_k c_{n_k} \to 0$ as $k \to \infty$, there is a $\xi > 0$ such that for all large k,

$$\left(1 + \frac{\varsigma}{2}\right)^2\left(1 - \frac{\epsilon_k c_{n_k}}{2}\right) > 1 + \xi.$$

In addition, by the way the sequence $\{n_k\}$ was selected in the proof of Lemma 4 (and in the way in which we select it for this proof) we have

$$P[S_{n_k} > s_{n_k} t_{n_k}(1 + \tfrac{1}{2}\varsigma)]$$

$$\leq \exp\{-(1 + \xi)\log\log c^{2k}\}$$

$$= \frac{1}{(2k\log c)^{1+\xi}}.$$

However, $\sum_{k=1}^{\infty} 1/k^{1+\xi}$ converges, and this proves (3).

We now prove (4). Let $\{n_k\}$, δ, c be as selected in the proof of (3). Since $1 > \epsilon > \delta > 0$, and since $c > 1$, there is a positive integer r such that

(8) $$1 > (1 - \delta)\left(1 - \frac{1}{c^{2r}}\right)^{1/2} - \frac{2}{c^r} > 1 - \epsilon.$$

Keeping r fixed we now define a sequence of integers $\{m_k\}$ by $m_k = n_{rk}$. We observe that

$$\frac{s_{m_{k+1}}}{s_{m_k}} = \frac{s_{n_{rk+r}}}{s_{n_{rk}}}$$

$$= \prod_{j=1}^{r}\left(\frac{s_{n_{kr+i}}}{s_{n_{kr+i-1}}}\right) \sim c^r \qquad \text{(as } k \to \infty).$$

Hence $\log(s^2_{m_k}/s^2_{m_{k-1}}) \sim \log c^{2r}$. We easily observe, by means of Lemma

1 in Section 5.3, that

$$\log s^2_{m_k} = \log \prod_{j=1}^{k} \left(\frac{s^2_{m_j}}{s^2_{m_{j-1}}} \right)$$

$$= k \left\{ \frac{1}{k} \sum_{j=1}^{k} \log \left(\frac{s^2_{m_j}}{s^2_{m_{j-1}}} \right) \right\}$$

$$\sim k \log c^{2r} = \log c^{2kr},$$

or

(9) $$\log s^2_{m_k} \sim \log c^{2kr} \qquad (\text{as } k \to \infty).$$

Denote

$$u_k^2 = s^2_{m_k} - s^2_{m_{k-1}} = s^2_{m_k} \left(1 - \frac{s^2_{m_{k-1}}}{s^2_{m_k}} \right)$$

$$\sim s^2_{m_k} \left(1 - \frac{1}{c^{2r}} \right),$$

and denote

$$v_k^2 = (2 \log \log u_k^2) \sim 2 \log \left(\log s^2_{m_k} + \log \left(1 - \frac{1}{c^{2r}} \right) \right)$$

$$\sim 2 \log \log s^2_{m_k} = t^2_{m_k}.$$

Consider the event

$$A_k = [S_{m_k} - S_{m_{k-1}} > (1 - \delta) u_k v_k].$$

We now prove that $P[A_k \text{ i.o.}] = 1$. We first note that the random variables $\{S_{m_k} - S_{m_{k-1}}, k = 1, 2, \cdots\}$ are independent, and hence the events $\{A_k\}$ are independent. By the Borel lemma (Theorem 2(b) in Section 3.3) we need only prove that $\sum P(A_k) = \infty$. Since $0 < \delta < 1$, if we let $\epsilon_k = (1 - \delta) v_k$, then $\epsilon_k \to \infty$ as $k \to \infty$. Let

$$c_k = \max \{\text{ess sup} \mid X_n \mid /u_k \mid m_{k-1} < n \leq m_k\}.$$

By hypothesis and the definition of u_k, we have for some

$$j_k \in \{m_{k-1} + 1, \cdots, m_k\},$$

$$c_k = \text{ess sup} \mid X_{j_k} \mid /u_k$$

$$\sim \left(1 - \frac{1}{c^{2r}} \right)^{-1/2} \text{ess sup} \mid X_{j_k} \mid /s_{m_k}$$

$$= \left(1 - \frac{1}{c^{2r}} \right)^{-1/2} \frac{s_{j_k}}{s_{m_k}} \text{ess sup} \mid X_{j_k} \mid /s_{j_k} \to 0$$

as $k \to \infty$. Next denote $\gamma = 1/(1 - \delta) - 1$, or $1 + \gamma = 1/(1 - \delta)$. For k sufficiently large we have by Theorem 2 in Section 5.5 and by (9),

$$
\begin{aligned}
P(A_k) &> \exp\{-\tfrac{1}{2}(1 - \delta)^2 v_k^2(1 + \gamma)\} \\
&= \exp\{-(1 - \delta) \log \log u_k^2\} \\
&= \exp\{-\log (\log u_k^2)^{1-\delta}\} \\
&= 1/(\log u_k^2)^{1-\delta} \\
&\sim 1/(\log c^{2kr})^{1-\delta} \\
&= 1/(2k \log c^r)^{1-\delta}.
\end{aligned}
$$

Since $\sum 1/k^{1-\delta}$ diverges, then $\sum P(A_k)$ diverges and $P[A_k \text{ i.o.}] = 1$. Next denote

$$
B_k = [\,|\,S_{m_{k-1}}\,| \leq 2s_{m_{k-1}}l_{m_{k-1}}].
$$

Since $0 < \delta < 1$, then in our proof of (3) we achieved a proof of

$$
P[B_k^c \text{ i.o.}] = 0,
$$

since $\{m_k\} \subset \{n_k\}$. Hence the probability that all but a finite number of B_k's occur is 1, from which we obtain

(10) $$P[A_kB_k \text{ i.o.}] = 1.$$

However,

$$
\begin{aligned}
A_kB_k &= [S_{m_k} - S_{m_{k-1}} > (1 - \delta)u_kv_k][\,|\,S_{m_{k-1}}\,| \leq 2s_{m_{k-1}}l_{m_{k-1}}] \\
&\subset [S_{m_k} > (1 - \delta)u_kv_k + S_{m_{k-1}}][S_{m_{k-1}} \geq -2s_{m_{k-1}}l_{m_{k-1}}] \\
&\subset [S_{m_k} > (1 - \delta)u_kv_k - 2s_{m_{k-1}}l_{m_{k-1}}].
\end{aligned}
$$

Since $u_k^2 \sim s_{m_k}^2(1 - 1/c^{2r})$, $v_k \sim l_{m_k}$ and (by Lemma 5)

$$
s_{m_{k-1}}l_{m_{k-1}} \sim (1/c^r)s_{m_k}l_{m_k},
$$

we have

$$
(1 - \delta)u_kv_k - 2s_{m_{k-1}}l_{m_{k-1}} \sim (1 - \delta)s_{m_k}l_{m_k}\left(1 - \frac{1}{c^{2r}}\right)^{1/2} - 2s_{m_{k-1}}l_{m_{k-1}}
$$

$$
\sim s_{m_k}l_{m_k}\left\{(1 - \delta)\left(1 - \frac{1}{c^{2r}}\right)^{1/2} - \frac{2}{c^r}\right\}.
$$

Then, by the way r was selected in order to satisfy (8) we have for all sufficiently large k,

(11) $$A_k B_k \subset [S_{m_k} > (1 - \epsilon) s_{m_k} t_{m_k}].$$

From (10) and (11) we may conclude that

$$P[S_{m_k} > (1 - \epsilon) s_{m_k} t_{m_k} \text{ i.o.}] = 1,$$

which in turn implies (4). Q.E.D.

There is a vast literature on the law of the iterated logarithm. Two papers in particular should be mentioned here. In 1937, J. Marcinkiewicz and A. Zygmund published an example which shows that the hypothesis $d_n = o(s_n/\sqrt{\log \log s_n^2})$ cannot be replaced by $d_n = O(s_n/\sqrt{\log \log s_n^2})$. [Reference: J. Marcinkiewicz and A. Zygmund, "Remarque sur la loi du logarithme iteré," *Fundamenta Mathematicae*, Vol. **29** (1937), pages 215–222.] P. Hartman and A. Wintner published a paper in 1941 in which they prove the law of the iterated logarithm when $\{X_n\}$ are unbounded but are identically distributed with common finite second moment. [Reference: P. Hartman and A. Wintner, "On the law of the iterated logarithm," *Am. J. of Math.*, Vol. **63** (1941), pages 169–176.]

EXERCISES

1. Prove that if $\{A_k\}$ and $\{B_k\}$ are sequences of events such that

$$P[A_k \text{ i.o.}] = 1 \quad \text{and} \quad P[B_k^c \text{ i.o.}] = 0,$$

then $P[A_k B_k \text{ i.o.}] = 1$.

2. Prove that if $\{X_n\}$ is a sequence of random variables, if $\lambda_n > 0$ and $\rho_n > 0$ for all n, if $\lambda_n \sim \rho_n$, and if $P[X_n > \lambda_n \text{ i.o.}] = 1$, then for $0 < \xi < 1$, $P[X_n > \xi \rho_n \text{ i.o.}] = 1$.

3. Prove that if μ is a median of the random variable X, then $-\mu$ is a median for $-X$.

4. If X is a random variable such that

$$EX = 0 \quad \text{and} \quad \text{ess sup} \, |X| = c < \infty,$$

then prove that $|E(X^n)| \leq c^n$ for all n.

5. Let $\{A_n\}$ and $\{B_n\}$ be sequences of events such that $A_n \subset B_n$ for all n. Prove that $[A_n \text{ i.o.}] \subset [B_n \text{ i.o.}]$.

6. Give two proofs of the following: a particle (or your fortune) at time 0 is at point 0. At time 1 it can move from 0 to 1 with probability p and from 0 to -1 with probability $1 - p = q$. If at time k ($k = 1$, 2, 3, $\cdots$) it is at point n ($n = 0, \pm 1, \pm 2, \cdots$), then independent of its past history it can move at time $k + 1$ to point $n + 1$ with probability p and to a point $n - 1$ with probability $q = 1 - p$. Prove that the particle (or your fortune) returns to 0 infinitely often with probability 1 *if and only if* $p = \frac{1}{2}$ (and with probability 0 if and only if $p \neq \frac{1}{2}$).

CHAPTER 6

The Central Limit Theorem

6.1. Infinitely Divisible Distributions

Among some probabilists the expression "the central limit theorem" refers to a body of theorems, each of which deals with conditions under which distribution functions of sums of random variables converge. More specifically, for every positive integer n, $\{X_{n,1}, \cdots, X_{n,k_n}\}$ will denote k_n independent random variables which obey these properties: (a) $k_n \to \infty$ as $n \to \infty$, and (b) for every sequence of integers $\{j_n\}$ for which $1 \leq j_n \leq k_n$ for every n, then $X_{n,j_n} \xrightarrow{P} 0$. Such a system of random variables will be called an *infinitesimal system*. We shall be concerned with conditions under which the distribution function of $S_n = \sum_{k=1}^{k_n} X_{n,k}$ converges as $n \to \infty$ and, in addition, shall be concerned with the distinguishing features of the class of limit distributions of such sums. It will turn out that the limit distributions have a special property, called "infinite divisibility." This first section of this chapter is accordingly devoted to an introduction to the infinitely divisible distribution functions.

Definition: A distribution function F is said to be infinitely divisible if for every positive integer n there is a distribution function F_n such that F is the n-fold convolution of F_n with itself, that is, $F = F_n * \cdots * F_n$ (n times).

An equivalent definition is that a distribution function F with characteristic function f is infinitely divisible if for every positive integer n there is a characteristic function f_n such that $f(u) = (f_n(u))^n$ for all u. Still another way of expressing this is as follows: F is an infinitely divisible distribution function if there is a probability space $(\Omega, \mathcal{Q}, P)$ such that for every integer n there is a random variable X_n and n independent identically distributed random variables $X_{n,1}, \cdots, X_{n,n}$ such that $X_n = X_{n,1} + \cdots + X_{n,n}$ and the distribution function of X_n is F. We shall say that f is an infinitely divisible characteristic function or X is an infinitely divisible random variable if the corresponding distribution function is infinitely divisible.

The principal tool used in this chapter is the characteristic function (see Section 2.5 and sections following it). In particular, explicit and implicit use will be made of the continuity theorem and the uniqueness theorem. The following are the two most important examples of infinitely divisible distribution functions.

(a) **The normal distribution,** $\mathfrak{N}(\mu, \sigma^2)$. In this case the density is

$$f(x) = (2\pi\sigma^2)^{-1/2} \exp\left\{-\frac{(x - \mu)^2}{2\sigma^2}\right\}$$

for $-\infty < x < \infty$, where μ, σ^2 are constants, $-\infty < \mu < \infty$, $\sigma^2 > 0$. The characteristic function of this distribution function is (see Section 2.5)

$$\varphi(u) = \exp\{i\mu u - \tfrac{1}{2}\sigma^2 u^2\}.$$

This characteristic function is clearly infinitely divisible, since

$$\varphi_n(u) = \exp\left\{i\frac{\mu}{n}u - \frac{1}{2}\frac{\sigma^2}{n}u^2\right\}$$

is a characteristic function of a normal distribution with expectation μ/n and variance σ^2/n, that is, of $\mathfrak{N}(\mu/n, \sigma^2/n)$, and $\varphi(u) = (\varphi_n(u))^n$.

(b) **The Poisson distribution,** $\mathcal{P}(\lambda)$. In this case a probability equal to $e^{-\lambda}\lambda^n/n!$ is assigned to each nonnegative integer n, and its characteristic function was computed previously as

$$\varphi(u) = \exp\{\lambda(e^{iu} - 1)\}.$$

This characteristic function is clearly infinitely divisible since

$$\varphi_n(u) = \exp\{(\lambda/n)(e^{iu} - 1)\}$$

is the characteristic function of the Poisson distribution with expectation λ/n, that is, of $\mathcal{P}(\lambda/n)$ and $\varphi(u) = (\varphi_n(u))^n$.

An easy example of a distribution function that is *not* infinitely divisible is

$$F(x) = \begin{cases} 0 & \text{if } x < 0 \\ \frac{1}{2} & \text{if } 0 \le x < 1 \\ 1 & \text{if } x \ge 1. \end{cases}$$

Theorem 1. If F is an infinitely divisible distribution function, then so is $G(x) = 1 - F(-x - 0)$.

Proof: If f, g are characteristic functions of F, G, respectively, then from the proof of Proposition 5 in Section 2.5, $g(u) = \overline{f(u)}$. Hence if $f_n(u)$ is a characteristic function such that $(f_n(u))^n = f(u)$, then $g(u) = (\overline{f_n(u)})^n$. Since $\overline{f_n(u)}$ is a characteristic function, we may conclude that G is infinitely divisible.

Theorem 1 is equivalent to the assertion that if $f(u)$ is an infinitely divisible characteristic function, then so is $\overline{f(u)}$.

Theorem 2. If X and Y are independent random variables with infinitely divisible distribution functions, then $X + Y$ has an infinitely divisible distribution function.

Proof: Let f, g be characteristic functions of X, Y, respectively. For each positive integer n let f_n and g_n be characteristic functions which satisfy $f(u) = (f_n(u))^n$ and $g(u) = (g_n(u))^n$. Then

$$f(u)g(u) = (f_n(u)g_n(u))^n.$$

Since $f_n(u)g_n(u)$ is a characteristic function, and since $f(u)g(u)$ is the characteristic function of $X + Y$, the conclusion of the theorem is established.

Theorem 3. If F is an infinitely divisible distribution function, then its characteristic function f never vanishes.

Proof: By Theorems 1 and 2, $|f(u)|^2$ is an infinitely divisible characteristic function. Let $f_n(u)$ be a characteristic function such that $(f_n(u))^n = f(u)$. Then $|f(u)|^{2/n} = |f_n(u)|^2$ is a characteristic function. But

$$|f(u)|^{2/n} = |f_n(u)|^2 \to \begin{cases} 1 & \text{over} \quad \{u \mid f(u) \neq 0\} \\ 0 & \text{over} \quad \{u \mid f(u) = 0\}. \end{cases}$$

Since f is continuous and $f(0) = 1$, then $f(u) \neq 0$ in a neighborhood of 0. Hence $\lim_{n \to \infty} |f(u)|^{2/n} = 1$ in a neighborhood of 0. By the continuity theorem, $\lim_{n \to \infty} |f(u)|^{2/n}$ is a characteristic function. Since its only values are 0 and 1, and since all characteristic functions are continuous, it cannot be zero at any value of u. Q.E.D.

This last theorem, whose proof depended on the first two theorems, is crucial in that it allows us to make use of logarithms of nonvanishing characteristic functions as defined in Section 4.3.

Theorem 4. If F is an infinitely divisible distribution function there is (for fixed n) only one distribution function F_n such that

$$F = F_n * \cdots * F_n \qquad (n \text{ times}).$$

Proof: Let f be the characteristic function of F, and let f_n be any characteristic function such that $(f_n(u))^n = f(u)$. By Theorem 1 in Section 4.3, there exists one and only one continuous, real-valued function φ_n such that $\varphi_n(0) = 0$ and $f_n(u) = |f_n(u)| \exp\{i\varphi_n(u)\}$, and one and only one such φ such that $f(u) = |f(u)| \exp\{i\varphi(u)\}$. From $f(u) = (f_n(u))^n$ it follows that $|f_n(u)|^n = |f(u)|$, and since $f(u) \neq 0$ for all u, then $\exp\{in\varphi_n(u)\} = \exp\{i\varphi(u)\}$. Now each side of this last equation satisfies the hypothesis of Theorem 1 in Section 4.3, and thus by the uniqueness statement in that theorem it follows that $n\varphi_n(u) = \varphi(u)$. Thus, for every $f_n(u)$ for which $(f_n(u))^n = f(u)$,

$$|f_n(u)| = |f(u)|^{1/n}$$

and

$$\varphi_n(u) = \frac{\varphi(u)}{n};$$

that is, there is just one $f_n(u)$.

Theorem 5. If $\{F_n\}$ is a sequence of infinitely divisible distribution functions, and if $F_n \overset{c}{\to} F$, then F is infinitely divisible.

Proof: For every n, let $F_{n,k}$ be *the* distribution function such that $F_n = F_{n,k} * \cdots * F_{n,k}$ (k times), and let $f_{n,k}$ be the characteristic function of $F_{n,k}$. By the Helly compactness theorem, for each fixed k there is a subsequence $\{F_{n_j,k}\}$ of $\{F_{n,k}\}$ such that, for some bounded nondecreasing function G_k, $F_{n_j,k} \overset{w}{\to} G_k$ as $j \to \infty$. We now show that G_k is a distribution function. If it were not, then there is a number $c \in (0, 1)$ such that $G_k(x) < c$ for every x or $G_k(x) > c$ for every x. Let us consider only the case $G_k(x) < c < 1$ for every x. Let

$$0 < \epsilon < 1 - c,$$

and let $K > 0$ be so large that $F(K) > 1 - \epsilon^k$ and such that F is continuous at K and G_k is continuous at K/k. Let $X_{j,1}, \cdots, X_{j,k}$ be k independent, identically distributed random variables with common distribution function $F_{n_j,k}$. Hence F_{n_j} is the distribution function of $X_j = X_{j,1} + \cdots + X_{j,k}$. Since

$$\bigcap_{i=1}^{k} \left[X_{j,i} > \frac{K}{k} \right] \subset [X_j > K],$$

we have, by taking probabilities, that

$$\left(1 - F_{n_j,k}\left(\frac{K}{k}\right) \right)^k \leq 1 - F_{n_j}(K),$$

and taking limits as $j \to \infty$ we have $(1 - G_k(K/k))^k \leq \epsilon^k$. Hence $G_k(K/k) \geq 1 - \epsilon > c$, which gives us a contradiction, and therefore G_k is a distribution function. Thus for every r, the distribution function of $X_{j,r}$ converges to the distribution function G_k of a random variable Y_r, where $Y_1, \cdots, Y_k$ are k independent random variables with common distribution function G_k. The joint distribution function of $(X_{j,1}, \cdots, X_{j,k})$ converges (as $j \to \infty$) to the joint distribution function of $(Y_1, \cdots, Y_k)$. By Theorem 3 of Section 4.2, the distribution function of $X_{j,1} + \cdots + X_{j,k}$ converges (as $j \to \infty$) to that of $Y_1 + \cdots + Y_k$. The distribution function of $X_{j,1} + \cdots + X_{j,k}$ is F_{n_j}, however, and the distribution function of $Y_1 + \cdots + Y_k$ is $G_k^{*k} = G_k * \cdots * G_k$ (k times). But also $F_{n_j} \overset{c}{\to} F$. Hence, for every k there is a distribution function G_k such that $F = G_k^{*k}$; that is, F is infinitely divisible. Q.E.D.

The following interesting theorem is worth keeping in mind while dealing with infinitely divisible distribution functions.

Theorem 6. Let $\{Y, X_1, X_2, \cdots\}$ be independent random variables, where the distribution of Y is Poisson with expectation $\lambda > 0$ and the $\{X_n\}$ are identically distributed with common characteristic function f. Let $Z = X_1 + \cdots + X_Y$. Then the distribution function of Z is infinitely divisible, and its characteristic function is

$$f_Z(u) = \exp(\lambda(f(u) - 1)).$$

Proof: We compute

$$f_Z(u) = E(e^{iuZ}) = \sum_{n=0}^{\infty} E(e^{iuZ} I_{[Y=n]})$$

$$= \sum_{n=0}^{\infty} E(I_{[Y=n]} \exp\{iu(X_1 + \cdots + X_n)\}).$$

Since $I_{[Y=n]}$ and $\exp\{iu(X_1 + \cdots + X_n)\}$ are independent, we have

$$f_Z(u) = \sum_{n=0}^{\infty} E(\exp\{iu(X_1 + \cdots + X_n)\})P[Y = n]$$

$$= \sum_{n=0}^{\infty} f^n(u)e^{-\lambda}\lambda^n/n! = \exp\{\lambda(f(u) - 1)\},$$

which proves the theorem.

EXERCISES

1. Let

$$F(x) = \begin{cases} 0 & \text{if } x < 0 \\ \frac{1}{2} & \text{if } 0 \le x < 1 \\ 1 & \text{if } x \ge 1. \end{cases}$$

Prove that F is *not* infinitely divisible.

2. Prove that if f is an infinitely divisible characteristic function, then so is $f^c(u) = \exp\{c \log f(u)\}$ for every $c > 0$.

3. Find an example of a characteristic function that does not vanish but is not infinitely divisible.

4. Prove that each of the asserted equivalent definitions of infinite divisibility that follows the definition is indeed equivalent to the definition given.

5. Prove that if $f(u)$ is any characteristic function, and if $c > 0$, then $\exp\{c(f(u) - 1)\}$ is a characteristic function which is infinitely divisible.

6.2. Canonical Representation of Infinitely Divisible Characteristic Functions

The purpose of this section is to derive the Lévy-Khinchine representation of the characteristic function of an infinitely divisible distribution function. This is accomplished in Theorem 1 in which the uniqueness of such a representation is also obtained.

Lemma 1. If f is the characteristic function of an infinitely divisible distribution function, and if f_n is the characteristic function for which $f(u) = (f_n(u))^n$, then $n(f_n(u) - 1) \to \log f(u)$ as $n \to \infty$ uniformly over every bounded interval.

Proof: Let $\varphi(u)$ be that unique real-valued continuous function for which $\varphi(o) = 0$ and such that $f(u) = |f(u)| \exp\{i\varphi(u)\}$. In Theorem 4 in Section 6.1 we proved that there was only one characteristic function $f_n(u)$ for which $(f_n(u))^n = f(u)$, and

$$f_n(u) = |f(u)|^{1/n} \exp\{i\varphi(u)/n\}.$$

Thus

$$f_n(u) = \exp\{(\log|f(u)| + i\varphi(u))/n)\},$$

from which we obtain

$$n(f_n(u) - 1) = n\{1 + (1/n)(\log|f(u)| + i\varphi(u)) + (1/n)o(1) - 1\}.$$

Since $o(1)$ is uniform over every bounded interval,

$$n(f_n(u) - 1) \to \log|f(u)| + i\varphi(u)$$

uniformly over every bounded interval. Q.E.D.

Lemma 2. If $A(x, u)$ is defined for $x \in (-\infty, 0) \cup (0, \infty)$ and $u \in (-\infty, +\infty)$ by

$$A(x, u) = \left(e^{ixu} - 1 - \frac{ixu}{1 + x^2}\right)\frac{1 + x^2}{x^2},$$

then $\lim_{x \to 0} A(x, u) = -u^2/2$.

Proof: This lemma is easily obtained by applying L'Hospital's rule twice.

Theorem 1. A function f is the characteristic function of an infinitely divisible distribution function if and only if there exists a real number γ and a bounded nondecreasing function G defined over $(-\infty, +\infty)$ such that

$$(1) \quad f(u) = \exp\left\{i\gamma u + \int_{-\infty}^{\infty}\left(e^{iux} - 1 - \frac{iux}{1 + x^2}\right)\frac{1 + x^2}{x^2}\, dG(x)\right\},$$

where the integrand is defined at $x = 0$ by continuity (see Lemma 2) to be $-u^2/2$. This representation is unique.

Proof: Suppose F is an infinitely divisible distribution function with characteristic function f. Let f_n be the characteristic function such that $f(u) = (f_n(u))^n$, $n = 1, 2, \cdots$, and let F_n be the distribution function of which f_n is the characteristic function. By Lemma 1,

$$(2) \quad n(f_n(u) - 1) = n\int_{-\infty}^{\infty}(e^{iux} - 1)\, dF_n(x) \to \log f(u) \qquad \text{as } n \to \infty$$

uniformly over every bounded interval. Let us denote

$$G_n(u) = n\int_{-\infty}^{u}\frac{x^2}{1 + x^2}\, dF_n(x)$$

and

$$I_n(u) = \int_{-\infty}^{\infty}(e^{iux} - 1)\frac{1 + x^2}{x^2}\, dG_n(x).$$

Then (2) asserts that $I_n(u) \to \log f(u)$ uniformly in every bounded interval. Then, if Re denotes "the real part of," we have

$$\text{Re } I_n(u) = \int_{-\infty}^{\infty} (\cos ux - 1) \frac{1 + x^2}{x^2} dG_n(x) \to \log |f(u)|$$

$$\text{as } n \to \infty.$$

Next we observe that $G_n(-\infty) = 0$. We now wish to show that there exists a bounded nondecreasing function G and a subsequence $\{G_{n_k}\}$ of $\{G_n\}$ such that $G_{n_k} \xrightarrow{c} G$ as $k \to \infty$. In order to do this we must show, by Theorem 4 in Section 4.1, that $\{G_n(+\infty)\}$ are bounded and that

$$\int_{|x| \geq T} dG_n(x) \to 0 \qquad \text{as } T \to \infty$$

uniformly in n. Let us denote

$$A_n = \int_{|x| \leq 1} dG_n(x),$$

$$B_n = \int_{|x| > 1} dG_n(x),$$

and

$$C_n = A_n + B_n = \int_{-\infty}^{\infty} dG_n(x) = G_n(+\infty).$$

Because of Re $I_n(u) \to \log |f(u)|$ as noted above, we have that over a bounded interval J, for arbitrary $\epsilon > 0$, and for sufficiently large n,

(3) $\qquad -\log |f(u)| + \epsilon \geq \int_{|x| \leq 1} (1 - \cos ux) \frac{1 + x^2}{x^2} dG_n(x)$

and

(4) $\qquad -\log |f(u)| + \epsilon \geq \int_{|x| > 1} (1 - \cos ux) \frac{1 + x^2}{x^2} dG_n(x).$

We consider (3). Clearly $(1 - \cos ux)/(ux)^2 \to \frac{1}{2}$ as $ux \to 0$. If we let $u = 1$, then $(1 - \cos x)/x^2 > 0$ and is bounded away from 0 in $[-1, 1]$;

thus for some positive constant K_0, we have

(5) $\qquad -\log |f(1)| + \epsilon \geq \int_{|x|\leq 1} K_0(1 + x^2)\, dG_n(x) \geq K_0 A_n.$

In (4) let us integrate both sides over the interval $J = [0, 2]$, divide by 2, and use Fubini's theorem to obtain

$$\frac{1}{2} \int_0^2 (-\log |f(u)|)\, du + \epsilon$$

$$\geq \int_{|x|>1} \left(\frac{1}{2} \int_0^2 (1 - \cos ux)\, du \right) \frac{1 + x^2}{x^2}\, dG_n(x)$$

$$\geq \int_{|x|>1} \left(1 - \frac{\sin 2x}{2x} \right) dG_n(x) > K_1 B_n$$

for some $K_1 > 0$. Hence $\{A_n\}$ and $\{B_n\}$ are both bounded, and so is

$$C_n = \int_{-\infty}^\infty dG_n(x).$$

Next we need to show that for arbitrary $\zeta > 0$ there is a T such that for all sufficiently large n,

$$\int_{|x|\geq T} dG_n(x) < \zeta.$$

As before, we integrate both sides of (4) over the interval $J = [0, 2/T]$, divide by $2/T$, and use Fubini's theorem to obtain for all $T > 0$,

$$\frac{T}{2} \int_0^{2/T} (-\log |f(u)|)\, du + \epsilon \geq \int_{|x|\geq T} \left(1 - \frac{\sin (2x/T)}{2x/T} \right) dG_n(x),$$

where we make use of the fact that (4) would continue to hold if the integral over $\{|x| > 1\}$ were replaced by the integral over $\{|x| > T\}$. For all $|x| \geq T$ or $|x|/T \geq 1$ there is a $K_2 > 0$ such that

$$1 - (\sin (2x/T))/(2x/T) \geq K_2,$$

and hence

$$\frac{T}{2} \int_0^{2/T} (-\log |f(u)|)\, du + \epsilon \geq K_2 \int_{|x|\geq T} dG_n(x)$$

for all sufficiently large n and independently of T. Now since $|f(0)| = 1$ and $\log |f(u)|$ is continuous at 0, we have for sufficiently large T,

$$\frac{T}{2} \int_0^{2/T} (-\log |f(u)|) \, du < \epsilon.$$

Hence for almost all n (and therefore for all n) and sufficiently large T,

$$\int_{|x| \geq T} dG_n(x) < \frac{2\epsilon}{K_2}.$$

Thus by Theorem 4 in Section 4.1 there do exist subsequences of $\{G_n\}$ which do converge completely. Let $\{G_{n_k}\}$ be such a subsequence, and let G be such that $G_{n_k} \xrightarrow{c} G$. For fixed u, $A(x, u)$ as defined in Lemma 2 is a bounded, continuous function in x. By the Helly-Bray theorem

$$\int_{-\infty}^{\infty} A(x, u) \, dG_{n_k}(x) \to \int_{-\infty}^{\infty} A(x, u) \, dG(x) \qquad \text{as} \quad k \to \infty.$$

Denote

$$\gamma_{n_k} = \int \frac{1}{x} dG_{n_k}(x) = n_k \int \frac{x}{1 + x^2} dF_{n_k}(x).$$

Then

$$I_{n_k}(u) = \int A(x, u) \, dG_{n_k}(x) + iu\gamma_{n_k},$$

and hence there is a number γ such that $\gamma_{n_k} \to \gamma$ as $k \to \infty$. Since

$$I_{n_k}(u) \to \log f(u) \qquad \text{as} \quad k \to \infty$$

we have

$$f(u) = \exp \left\{ i\gamma u + \int A(x, u) \, dG(x) \right\},$$

which verifies that f can be represented as it is in (1).

Conversely, let us suppose that f is a function as is defined in (1); we shall show that it is the characteristic function of an infinitely divisible distribution function.

Case a: Suppose

$$0 < \lambda = \int_{-\infty}^{-0} + \int_{+0}^{\infty} \frac{1 + x^2}{x^2} dG(x) < \infty.$$

Let $I = (-\infty, 0) \cup (0, \infty)$, let

$$H(x) = \frac{1}{\lambda} \int_{I \cap (-\infty, x)} \frac{1 + x^2}{x^2} \, dG(x),$$

and let $\sigma^2 = G(+0) - G(-0)$. Then H is a distribution function, and if $h(u)$ is its characteristic function, then

$$f(u) = \exp\{i\gamma'u - \tfrac{1}{2}\sigma^2 u^2 + \lambda(h(u) - 1)\},$$

where

$$\gamma' = \gamma - \int_I (1/x) \, dG(x).$$

By Theorems 2 and 6 of Section 6.1, $f(u)$ is a characteristic function of an infinitely divisible distribution function. In case $\lambda = 0$, $\sigma^2 > 0$, then $f(u)$ is the characteristic function of the $\mathfrak{N}(\gamma, \sigma^2)$ distribution which was shown to be infinitely divisible in Section 6.1.

Case (b). In this case,

$$\int_I \frac{1 + x^2}{x^2} \, dG(x) = \infty.$$

Let $\{x_n\}$ be a sequence of positive numbers such that $x_n > x_{n+1}$ for all n and $x_n \to 0$ as $n \to \infty$. Let us denote

$$A_1 = (-\infty, -x_1] \cup [x_1, \infty)$$

and

$$A_n = (-x_{n-1}, -x_n] \cup [x_n, x_{n-1})$$

for $n = 2, 3, \cdots$, and finally let us require that

$$\lambda_n = \int_{A_n} \frac{1 + x^2}{x^2} \, dG(x) > 0$$

for every n. Note that $(-\infty, 0) \cup (0, \infty) = I = \cup_{n=1}^{\infty} A_n$. Let

$$F_n(x) = \lambda_n^{-1} \int_{A_n \cap (-\infty, x]} \frac{1 + x^2}{x^2} \, dG(x);$$

clearly F_n is a distribution function. Let

$$\{Z, X_{n,j}, Y_n, n = 1, 2, \cdots, j = 1, 2, \cdots\}$$

be independent random variables such that $\mathcal{L}(Z)$ is $\mathfrak{N}(0, \sigma^2)$, the distribution of $X_{n,j}$ is F_n, and $\mathcal{L}(Y_n)$ is $\mathcal{P}(\lambda_n)$. Let

$$X_n = \gamma + Z + \sum_{m=1}^{n} \left(X_{m,1} + \cdots + X_{m,Y_m} - \int_{A_m} x^{-1} \, dG(x) \right).$$

By Theorem 6 in Section 6.1, the characteristic function $f_n(u)$ of X_n is

$$f_n(u) = \exp \left\{ i\gamma u - \tfrac{1}{2}\sigma^2 u^2 + \int_{-\infty}^{-x_n} + \int_{x_n}^{\infty} A(x, u) \, dG(x) \right\},$$

where $A(x, u)$ is as defined in Lemma 2. Clearly, $f_n(u) \to f(u)$ as $n \to \infty$ at all u. Further, it is easy to verify that $f(u)$ as defined in (1) is continuous. Thus, by the continuity theorem, $f(u)$ is a characteristic function. By Theorem 5 in Section 6.1, $f(u)$ is the characteristic function of an infinitely divisible distribution function.

Before we prove the uniqueness of r and G, it should be remarked that

$$\log f(u) = i\gamma u + \int_{-\infty}^{\infty} A(x, u) \, dG(x).$$

This follows from the fact that $\log |f(u)|$ is continuous, from the fact that

$$\log |f(u)| - i\gamma u - \int_{-\infty}^{\infty} A(x, u) \, dG(x)$$

is continuous and vanishes at $u = 0$, and from Theorem 1 in Section 4.3. Now we prove uniqueness of γ and G. In order to prove uniqueness of γ we need only prove uniqueness of G. Let $\varphi(u)$ be defined by

$$-\varphi(u) = \int_{u-1}^{u+1} \log f(\tau) \, d\tau - 2 \log f(u).$$

Clearly, f uniquely determines φ. By Fubini's theorem we obtain

$$\varphi(u) = 2 \int_{-\infty}^{\infty} e^{iux} \left(1 - \frac{\sin x}{x} \right) \frac{1 + x^2}{x^2} \, dG(x).$$

Since one easily verifies that

$$\left(1 - \frac{\sin x}{x} \right) \frac{1 + x^2}{x^2} > 0$$

for all x, then if we define

$$\Phi(x) = 2 \int_{-\infty}^{x} \left(1 - \frac{\sin y}{y} \right) \frac{1 + y^2}{y^2} \, dG(y),$$

it follows that

$$\varphi(u) = \int_{-\infty}^{\infty} e^{iux} \, d\Phi(x).$$

(Note: Φ is nondecreasing since it is the integral of a nonnegative function. It is of finite total variation since $\varphi(0)$ as originally defined is finite.) Because the integrand defining Φ is *strictly positive*, it follows that Φ uniquely determines G. But by the uniqueness theorem (Theorem 1 in Section 2.5), φ uniquely determines Φ. Thus f uniquely determines G and also γ. Q.E.D.

Another representation of the characteristic function of an infinitely divisible distribution function is given by the following theorem.

Theorem 2. A function f is a characteristic function of an infinitely divisible distribution function if and only if there exist constants γ and $\sigma^2 \geq 0$ and a function M defined over $(-\infty, 0) \cup (0, \infty)$ which is nondecreasing over $(-\infty, 0)$ and over $(0, \infty)$ and satisfies $M(-\infty) = M(\infty) = 0$ and

$$\int_{-1}^{-0} + \int_{+0}^{+1} x^2 \, dM(x) < \infty,$$

and such that

(6) $\quad f(u) = \exp\left\{ i\gamma u - \frac{\sigma^2 u^2}{2} \right.$

$$\left. + \int_{-\infty}^{-0} + \int_{+0}^{\infty} \left(e^{iux} - 1 - \frac{iux}{1+x^2} \right) dM(x) \right\}.$$

This representation is unique.

This theorem is clearly equivalent to Theorem 1 by the relation

$$M(x) = \begin{cases} -\displaystyle\int_{x}^{\infty} \frac{1+y^2}{y^2} \, dG(y) & \text{if } x > 0 \\[4ex] \displaystyle\int_{-\infty}^{x} \frac{1+y^2}{y^2} \, dG(y) & \text{if } x < 0. \end{cases}$$

Formula (6) is referred to as the Lévy representation of $f(u)$.

EXERCISES

1. Prove that another unique representation of a characteristic function f of an infinitely divisible distribution function is

$$f(u) = \exp \left\{ i\gamma(\tau)u - \frac{\sigma^2 u^2}{2} + \int_{-\infty}^{-\tau} + \int_{\tau}^{\infty} (e^{iux} - 1) \, dM(x) \right.$$

$$\left. + \int_{-\tau}^{-0} + \int_{+0}^{\tau} (e^{iux} - 1 - iux) \, dM(x) \right\},$$

where M and $\sigma^2 \geq 0$ are as in Theorem 2, where $\gamma(\tau)$ depends on $\tau > 0$, and $\tau \in Cont\ M$.

2. In Theorem 1, let $G(x) = 0$ if $x < 0$ and $G(x) = 1$ if $x \geq 0$, and let $\gamma = 0$. Prove that f is the characteristic function of the $\mathfrak{N}(0, 1)$ distribution.

3. Find γ and G in (1) in order that $f(u) = \exp\{\lambda(e^{iu} - 1)\}$, that is, in order that $f(u)$ is the characteristic function of the Poisson distribution with expectation $\lambda > 0$.

4. Prove: If G and H are two bounded nondecreasing functions such that for some positive continuous function ρ,

$$G(x) = \int_{-\infty}^{x} \rho(\tau) \, dH(\tau), \quad \text{then} \quad H(x) = \int_{-\infty}^{x} \frac{1}{\rho(\tau)} \, dG(\tau).$$

6.3. Convergence of Infinitely Divisible Distribution Functions

We have shown in the representation theorem in Section 6.2 that the infinitely divisible distribution functions are in one-to-one correspondence in a special way with the set of all pairs $\{\gamma, G\}$, where γ is a real number and G is a bounded nondecreasing function. If $\{F_n\}$ is a sequence of infinitely divisible distribution functions with corresponding pair $\{\gamma_n, G_n)\}$, we know by Theorem 5 of Section 6.1 that if $F_n \xrightarrow{c} F$, then F is infinitely divisible with corresponding pair (γ, G). The question then arises whether this implies that $\gamma_n \to \gamma$ and $G_n \xrightarrow{c} G$ as $n \to \infty$. This question is completely answered in this section.

Theorem 1. Let $\{F_n\}$ be a sequence of infinitely divisible distribution functions with corresponding pairs $\{(\gamma_n, G_n)\}$. If $F_n \xrightarrow{c} F$, where (γ, G)

is the corresponding pair for F, then $\gamma_n \to \gamma$ and $G_n \xrightarrow{c} G$ as $n \to \infty$. Conversely, if there exist a constant γ and a bounded nondecreasing function G such that $\gamma_n \to \gamma$ and $G_n \xrightarrow{c} G$ as $n \to \infty$, then $F_n \xrightarrow{c} F$, where F is infinitely divisible with corresponding pair (γ, G).

Proof: Since $A(x, u)$ as defined in Lemma 2 in Section 6.2 is a bounded, continuous function in x, then by the Helly-Bray theorem, Theorem 1 in Section 6.2, and the continuity theorem, the converse is easily obtained. We now prove the theorem. Let f_n and f be the characteristic functions of F_n and F, respectively. Since $F_n \xrightarrow{c} F$, then $f_n(u) \to f(u)$ at each u. By Theorem 4 of Section 4.3, $\log f_n(u) \to \log f(u)$, that is,

$$(1) \qquad i\gamma_n u + \int_{-\infty}^{\infty} A(x, u)\, dG_n(x) \to i\gamma u + \int_{-\infty}^{\infty} A(x, u)\, dG(x)$$

as $n \to \infty$. (We noted already in the proof of Theorem 1 of Section 6.2 that these expressions *are* the logarithms of the infinitely divisible characteristic functions.) This implies that $\operatorname{Re} \log f_n(u) \to \operatorname{Re} \log f(u)$ as $n \to \infty$, or

$$\int (\cos ux - 1)\, \frac{1 + x^2}{x^2}\, dG_n(x) \to \int (\cos ux - 1)\, \frac{1 + x^2}{x^2}\, dG(x)$$

as $n \to \infty$. However, when this occurred previously in the proof of Theorem 1 in Section 6.2, we proved that this implied that $\{G_n\}$ are uniformly bounded and that every subsequence $\{G_{n_k}\}$ of $\{G_n\}$ which converges in the wide sense converges completely. Since the $\{G_n\}$ are uniformly bounded, then by the Helly compactness theorem there is a subsequence $\{G_{n_k}\}$ which converges in the wide sense, and hence by the above remark there is a bounded nondecreasing function G^* such that $G_{n_k} \xrightarrow{c} G^*$. Since for each u, $A(x, u)$ is a bounded, continuous function of x, then by the Helly-Bray theorem

$$(2) \qquad \int_{-\infty}^{\infty} A(x, u)\, dG_{n_k}(x) \to \int_{-\infty}^{\infty} A(x, u)\, dG^*(x).$$

Thus (1) and (2) imply that there is a real number γ^* such that $\gamma_{n_k} \to \gamma^*$, and consequently

$$i\gamma^* u + \int_{-\infty}^{\infty} A(x, u)\, dG^*(x) = i\gamma u + \int_{-\infty}^{\infty} A(x, u)\, dG(x).$$

By the uniqueness assertion of Theorem 1 in Section 6.2, we may conclude that $\gamma^* = \gamma$ and $G^* = G$. Since every subsequence of $\{G_n\}$ which converges in the wide sense converges completely to G, and since $\{G_n\}$ are uniformly bounded, we may conclude that $G_n \overset{c}{\to} G$, from which we also obtain $\gamma_n \to \gamma$ as $n \to \infty$. Q.E.D.

An analogous theorem is needed when the representation of infinitely divisible distribution functions is given by (γ, σ^2, M) of Theorem 2 in Section 6.2, where γ is the centering constant, σ^2 is the variance of the Gaussian or normal component, and M is the Poisson component intensity function.

Theorem 2. Let $\{F_n\}$ be a sequence of infinitely divisible distribution functions whose corresponding representations are determined by the sequence of triples $\{(\gamma_n, \sigma_n^2, M_n)\}$. If $F_n \overset{c}{\to} F$, where F uniquely determines (γ, σ^2, M), then

(a) $\quad \gamma_n \to \gamma$

(b) $\quad M_n \overset{w}{\to} M$, and

(c) $\quad \lim_{\epsilon \downarrow 0} \lim_{n \to \infty} \sup \left\{ \int_{-\epsilon}^{\epsilon} x^2 \, dM_n(x) + \sigma_n^2 \right\}$

$$= \lim_{\epsilon \downarrow 0} \lim_{n \to \infty} \inf \left\{ \int_{-\epsilon}^{\epsilon} x^2 \, dM_n(x) + \sigma_n^2 \right\} = \sigma^2.$$

Conversely, if there exists a triple (γ, σ^2, M) such that (a), (b), and (c) hold and M satisfies the conditions of Theorem 2 of Section 6.2 then $F_n \overset{c}{\to} F$, where F is an infinitely divisible distribution function determined by (γ, σ^2, M).

Proof: The necessity of conditions (a) and (b) clearly follows from Theorem 1. We now show that (c) is necessary. Let $\epsilon > 0$ be such that $\pm\epsilon$ are continuity points of all M_n and M. We denote

$$I_n(\epsilon) = G_n(+\epsilon) - G_n(-\epsilon)$$

$$= \int_{-\epsilon}^{\epsilon} \frac{x^2}{1 + x^2} \, dM_n(x) + \sigma_n^2,$$

and

$$I(\epsilon) = G(\epsilon) - G(-\epsilon) = \int_{-\epsilon}^{\epsilon} \frac{x^2}{1 + x^2} \, dM(x) + \sigma^2.$$

Theorem 1 implies that $I_n(\epsilon) \to I(\epsilon)$ as $n \to \infty$. Easily we observe

(3) $$\frac{1}{1+\epsilon^2} \int_{-\epsilon}^{\epsilon} x^2 \, dM_n(x) \leq \int_{-\epsilon}^{\epsilon} \frac{x^2}{1+x^2} \, dM_n(x)$$

$$\leq \int_{-\epsilon}^{\epsilon} x^2 \, dM_n(x)$$

$$\leq (1+\epsilon^2) \int_{-\epsilon}^{\epsilon} \frac{x^2}{1+x^2} \, dM_n(x),$$

and the same inequalities hold without the subscript n. Let us denote

$$J_n(\epsilon) = \int_{-\epsilon}^{\epsilon} x^2 \, dM_n(x) + \sigma_n{}^2.$$

Then by (3),

$$\frac{J_n(\epsilon)}{1+\epsilon^2} \leq I_n(\epsilon) \leq J_n(\epsilon).$$

Thus,

$$\frac{1}{1+\epsilon^2} \limsup_{n \to \infty} J_n(\epsilon) \leq I(\epsilon) \leq \limsup_{n \to \infty} J_n(\epsilon).$$

Since $\limsup_{n \to \infty} J_n(\epsilon)$ is monotone nonincreasing as ϵ decreases, we obtain

$$\lim_{\epsilon \downarrow 0} \limsup_{n \to \infty} J_n(\epsilon) = \lim_{\epsilon \downarrow 0} I(\epsilon) = \sigma^2.$$

In a similar manner, one obtains

$$\lim_{\epsilon \downarrow 0} \liminf_{n \to \infty} J_n(\epsilon) = \sigma^2,$$

which concludes the proof of the necessity of (c). We now prove that (a), (b), and (c) are sufficient. Accordingly we must show that the conditions of Theorem 1 are satisfied, namely, $\gamma_n \to \gamma$ and $G_n \xrightarrow{c} G$. The condition $\gamma_n \to \gamma$ is exactly condition (a) here. In order to prove $G_n \xrightarrow{c} G$ we first recall the relation between G and $\{\sigma^2, M\}$, namely,

$$\sigma^2 = G(+0) - G(-0),$$

and

$$
G(x) \; = \;
\begin{cases}
\displaystyle\int_{-\infty}^{x} \frac{t^2}{1 + t^2}\, dM(t) & \text{if } \; x < 0 \;\; \text{and} \;\; x \in Cont\ G, \\[4ex]
\displaystyle\int_{-\infty}^{x} \frac{t^2}{1 + t^2}\, dM(t) + \sigma^2 & \text{if } \; x > 0 \;\; \text{and} \;\; x \in Cont\ G.
\end{cases}
$$

Let $x < 0$ and $x \in Cont\ G$. Then by the Helly-Bray theorem,

$$
(4) \qquad G_n(x) = \int_{-\infty}^{x} \frac{t^2}{1 + t^2}\, dM_n(t) \to \int_{-\infty}^{x} \frac{t^2}{1 + t^2}\, dM(t) = G(x)
$$

as $n \to \infty$. Further $G_n(-\infty) \to G(-\infty)$ as $n \to \infty$. It remains to prove that $G_n(x) \to G(x)$ for $0 < x$, $x \in Cont\ G$, and $G_n(+\infty) \to G(+\infty)$ as $n \to \infty$. Now $\lim \sup_{n\to\infty} G_n(-\epsilon)$ and $\lim \inf_{n\to\infty} G_n(-\epsilon)$ are non-decreasing as ϵ decreases. Since by (4) we have for every $-\epsilon \in Cont\ G$,

$$
\lim_{n\to\infty} \sup G_n(-\epsilon) = \lim_{n\to\infty} \inf G_n(-\epsilon) = \lim_{n\to\infty} G_n(-\epsilon) = G(-\epsilon),
$$

then we obtain

$$
\lim_{\epsilon\downarrow 0} \lim_{n\to\infty} \sup G_n(-\epsilon) = \lim_{\epsilon\downarrow 0} \lim_{n\to\infty} \inf G_n(-\epsilon) = G(-0).
$$

We observe that

$$
G_n(+\epsilon) = G_n(-\epsilon) + \int_{-\epsilon}^{\epsilon} \frac{x^2}{1 + x^2}\, dM_n(x) + \sigma_n^2
$$

$$
\leq G_n(-\epsilon) + \int_{-\epsilon}^{\epsilon} x^2\, dM_n(x) + \sigma_n^2.
$$

Condition (c) implies that

$$
\lim_{\epsilon\downarrow 0} \lim_{n\to\infty} \sup G_n(+\epsilon) \leq G(-0) + \sigma^2 = G(+0).
$$

Similarly

$$
G_n(+\epsilon) \geq G_n(-\epsilon) + (1 + \epsilon^2)^{-1}\left(\int_{-\epsilon}^{\epsilon} x^2\, dM_n(x) + \sigma_n^2 \right),
$$

from which we obtain

$$\lim_{\epsilon \downarrow 0} \lim_{n \to \infty} \inf G_n(+\epsilon) \geq G(-0) + \sigma^2 = G(+0).$$

Now let $x > 0$, $x \in Cont\ G$, let $0 < \epsilon < x$, and let $\epsilon \in Cont\ G$. Then

$$\lim_{n \to \infty} \sup G_n(x) = \lim_{n \to \infty} \sup (G_n(+\epsilon) + (G_n(x) - G_n(\epsilon)))$$

$$= \lim_{\epsilon \downarrow 0} \lim_{n \to \infty} \sup \left\{ G_n(\epsilon) + \int_\epsilon^x \frac{t^2}{1 + t^2}\, dM_n(t) \right\}$$

$$\leq G(+0) + \lim_{\epsilon \downarrow 0} \lim_{n \to \infty} \sup \int_\epsilon^x \frac{t^2}{1 + t^2}\, dM_n(t)$$

$$= G(+0) + \lim_{\epsilon \downarrow 0} \lim \int_\epsilon^x \frac{t^2}{1 + t^2}\, dM(t)$$

$$= G(+0) + \lim_{\epsilon \downarrow 0} (G(x) - G(\epsilon)) = G(x).$$

Similarly, $\lim \inf_{n \to \infty} G_n(x) \geq G(x)$ for $0 < x \in Cont\ G$, and thus $G_n(x) \to G(x)$ at all $x \in Cont\ G$. We have yet to prove that $G_n(+\infty) \to G(+\infty)$. If $\pm\epsilon$ are continuity points of all M_n and M, let us write

$$G_n(+\infty) = \int_{-\infty}^{-\epsilon} + \int_\epsilon^\infty \frac{x^2}{1 + x^2}\, dM_n(x) + \int_{-\epsilon}^\epsilon \frac{x^2}{1 + x^2}\, dM_n(x) + \sigma_n^2.$$

Then by (b) we obtain

$$\lim_{n \to \infty} \sup G_n(+\infty) \leq \int_\infty^{-\epsilon} + \int_{-\epsilon}^\infty \frac{x^2}{1 + x^2}\, dM(x)$$

$$+ \lim_{n \to \infty} \sup \left\{ \int_{-\epsilon}^\epsilon \frac{x^2}{1 + x^2}\, dM_n(x) + \sigma_n^2 \right\}.$$

If we take the limit of both sides as $\epsilon \to 0$ we find that

$$\lim_{n \to \infty} \sup G_n(+\infty) \leq G(+\infty) - G(+0) + G(-0) + \sigma^2 = G(+\infty).$$

Similarly, $\lim \inf_{n \to \infty} G_n(+\infty) \geq G(+\infty)$, which concludes the proof of the theorem.

EXERCISES

1. Prove that if $\{G_n\}$ is a uniformly bounded sequence of nondecreasing functions over $(-\infty, +\infty)$, and if every subsequence of $\{G_n\}$ which converges in the wide sense converges completely to G, then $G_n \xrightarrow{c} G$.

2. Construct a sequence of infinitely divisible distribution functions which satisfies Theorem 2 but for which it is not true that $\sigma_n^2 \to \sigma^2$ as $n \to \infty$.

6.4. Infinitesimal Systems of Random Variables

The purpose of this section is to define an infinitesimal system of random variables. In addition a crucial lemma, given as Theorem 3, is proved; it will be used in the very next section.

In this and all subsequent sections

$$\mathfrak{X} \doteq \{\{X_{n,1}, \cdots, X_{n,k_n}\}\}$$

will denote a double sequence of random variables which are row-wise independent; more specifically, $n = 1, 2, \cdots, k_n \to \infty$ as $n \to \infty$, and for every n (which designates the row), the random variables $X_{n,1}, \cdots,$ X_{n,k_n} are independent. A simple example of such a system is obtained by taking a sequence of independent identically distributed random variables $\{X_n\}$ and taking $k_n = n$ and $X_{n,j} = X_j/\sqrt{n}$, $1 \leq j \leq n$.

Definition: $\mathfrak{X}$ is called an infinitesimal system if for every sequence of integers $\{j_n\}$ which satisfy $1 \leq j_n \leq k_n$ for all n we have $X_{n,j_n} \xrightarrow{P} 0$.

The next two propositions give equivalent definitions of an infinitesimal system.

Proposition 1. $\mathfrak{X}$ is an infinitesimal system if and only if

$$\max_{1 \leq j \leq k_n} P[\,|X_{nj}| \geq \epsilon\,] \to 0$$

as $n \to \infty$ for every $\epsilon > 0$.

Proof: Let $\epsilon > 0$. Suppose $\mathfrak{X}$ is an infinitesimal system. Let j_n be defined by

$$(1) \qquad P[\,|\,X_{n,j_n}\,|\geq \epsilon\,] = \max_{1\leq j\leq k_n} P[\,|\,X_{nj}\,|\geq \epsilon\,].$$

Since $X_{n,j_n} \xrightarrow{P} 0$, then the right side of (1) tends to zero as $n \to \infty$. Conversely, for every sequence $\{\,j_n\}$ for which $1 \leq j_n \leq n$, the left side of (1) will not exceed the right side. Thus the condition of the proposition implies that $X_{n,j_n} \xrightarrow{P} 0$. Q.E.D.

In this and subsequent sections we use the following notation:

F_{nj} denotes the distribution function of X_{nj},

f_{nj} denotes the characteristic function of X_{nj}, and

m_{nj} denotes a median of X_{nj}.

Proposition 2. $\mathfrak{X}$ is an infinitesimal system if and only if

$$(2) \qquad \max_{1\leq j\leq k_n} |\,f_{nj}(u) - 1\,| \to 0 \qquad \text{as} \quad n \to \infty$$

uniformly over every bounded interval.

Proof: If (2) holds, then for every sequence $\{\,j_n\}$ for which $1 \leq j_n \leq k_n$, we have $f_{nj_n}(u) \to 1$ as $n \to \infty$, which in turn is easily shown to imply that $X_{nj_n} \xrightarrow{P} 0$. Conversely, since $X_{nj_n} \xrightarrow{P} 0$ for every sequence $\{\,j_n\}$ it follows that

$$\max_{1\leq j\leq k_n} |\,f_{nj}(u) - 1\,| \to 0 \qquad \text{as} \quad n \to \infty.$$

In fact, this convergence is uniform over every bounded interval. If it were not, then there would exist an $\epsilon > 0$, a $K > 0$, an increasing sequence of positive integers r_n, and a sequence $\{u_n\}$ of numbers in $[-K, K]$ such that for every n

$$\max_{1\leq j\leq k_{r_n}} |\,f_{r_n,j}(u_n) - 1\,| \geq \epsilon.$$

Let $\{\,j_n\}$ be defined by

$$|\,f_{r_n,j_n}(u_n) - 1\,| = \max_{1\leq j\leq k_{r_n}} |\,f_{r_n,j}(u_n) - 1\,|.$$

But $X_{r_n,j_n} \xrightarrow{P} 0$, and hence by Theorem 3 in Section 4.3, $f_{r_n,j_n}(u) \to 1$ uniformly over every bounded interval. Therefore, for large values of n,

$| f_{r_n, j_n}(u) - 1 | < \epsilon$ for all $u \in [-K, K]$, which contradicts the denial of uniform convergence over $[-K, K]$. Q.E.D.

Lemma 1. Let X and Y be independent, identically distributed random variables with common distribution function F, and let ψ be defined by

$$\psi(x) = \begin{cases} x^2 & \text{if } |x| \leq 1 \\ 1 & \text{if } |x| > 1. \end{cases}$$

If m_X denotes a median of X, then

$$E\psi(X - Y) \geq \tfrac{1}{2} E\psi(X - m_X).$$

Proof: Without loss of generality we may take $m_X = 0$. Since $\psi(x) \geq 0$ for all x, we have

$$E\psi(X - Y) = \iint \psi(x - y)\, dF(x)\, dF(y)$$

$$\geq \iint_{\{x \geq 0, y \leq 0\}} \psi(x - y)\, dF(x)\, dF(y)$$

$$+ \iint_{\{y \geq 0, x < 0\}} \psi(x - y)\, dF(x)\, dF(y).$$

Over the quadrant $\{x \geq 0, y \leq 0\}$ we observe that $x - y \geq x \geq 0$ which implies that $\psi(x - y) \geq \psi(x) \geq 0$. Similarly, over $\{x < 0, y \geq 0\}$, $\psi(x - y) \geq \psi(x) \geq 0$. Thus

$$E\psi(X - Y) \geq \int_{\{y \leq 0\}} dF(y) \int_{\{x \geq 0\}} \psi(x)\, dF(x)$$

$$+ \int_{\{y \geq 0\}} dF(y) \int_{\{x < 0\}} \psi(x)\, dF(x)$$

$$\geq \tfrac{1}{2} \int \psi(x)\, dF(x) = \tfrac{1}{2} E\psi(X). \quad \text{Q.E.D.}$$

Theorem 1. Let $\mathfrak{X} = \{\{X_{n,1}, \cdots, X_{n,k_n}\}\}$ be a double sequence of random variables which are row-wise independent. A necessary and sufficient condition that there exist a sequence of constants $\{c_n\}$ such that $X_{n,1} + \cdots + X_{n,k_n} - c_n \xrightarrow{P} 0$ is that

(a) $\displaystyle\sum_{j=1}^{k_n} \int_{|x|>1} dF_{nj}(x + m_{nj}) \to 0 \qquad$ as $\quad n \to \infty$, and

(b) $\displaystyle\sum_{j=1}^{k_n} \int_{|x|\leq 1} x^2 \, dF_{nj}(x + m_{nj}) \to 0 \qquad$ as $\quad n \to \infty.$

Proof: We first prove that (a) and (b) are sufficient. Let us denote $X'_{nk} = X_{nk} - m_{nk}$, $F'_{nk}(x) = P[X'_{nk} \leq x] = F_{nk}(x + m_{nk})$, $X''_{nk} = X'_{nk}I_{[|X'_{nk}|\leq 1]}$, $c_n = \sum_{j=1}^{k_n} (m_{nj} + EX''_{nj})$, $X_n = \sum_{j=1}^{k_n} X_{nj}$, $X'_n = \sum_{j=1}^{k_n} X'_{nj}$, $X''_n = \sum_{j=1}^{k_n} X''_{nj}$, and $B_n = [X'_n = X''_n]$. It is clear that $\cap_{j=1}^{k_n}[X'_{nj} = X''_{nj}] \subset B_n$, from which we obtain

$$P(B_n{}^c) \leq \sum_{j=1}^{k_n} P[X'_{nj} \neq X''_{nj}] = \sum_{j=1}^{k_n} P[\,|X'_{nj}| > 1]$$

$$= \sum_{j=1}^{k_n} \int_{|x|>1} dF_{nj}(x + m_{nj}).$$

By (a) in the hypothesis, $P(B_n{}^c) \to 0$ and $P(B_n) \to 1$ as $n \to \infty$. We may write

$$P[\,|X_n - c_n| \geq \epsilon] = P([\,|X_n - c_n| \geq \epsilon]B_n)$$
$$+ P([\,|X_n - c_n| \geq \epsilon]B_n{}^c).$$

Using Chebishev's inequality we obtain

$$P([\,|X_n - c_n| \geq \epsilon]B_n) = P([\,|\sum_{j=1}^{k_n} (X''_{nj} - EX''_{nj})\,| \geq \epsilon]B_n)$$

$$\leq P[\,|\sum_{j=1}^{k_n} (X''_{nj} - EX''_{nj})\,| \geq \epsilon]$$

$$\leq \frac{1}{\epsilon^2} \sum_{j=1}^{k_n} \mathrm{Var}\,(X''_{nj}) \leq \frac{1}{\epsilon^2} \sum_{j=1}^{k_n} E(X''_{nj}{}^2)$$

$$= \frac{1}{\epsilon^2} \sum_{j=1}^{k_n} \int_{|x|\leq 1} x^2 \, dF_{nj}(x + m_{nj}) \to 0$$

as $n \to \infty$ by (b) in the hypothesis. Also

$$P([\,|X_n - c_n| \geq \epsilon]B_n{}^c) \leq P(B_n{}^c) \to 0 \qquad \text{as} \quad n \to \infty,$$

so

$$P[\,|X_n - c_n| \geq \epsilon] \to 0, \quad \text{that is} \quad X_n - c_n \overset{P}{\to} 0.$$

We now prove that conditions (a) and (b) are necessary. Accordingly we assume that there exists a sequence of constants $\{c_n\}$ such that $X_n - c_n \overset{P}{\to} 0$. Hence $\mathcal{L}(X_n - c_n) \to \mathcal{L}(0)$, and thus we obtain

$$\exp\,(-ic_n u) \prod_{j=1}^{k_n} f_{nj}(u) \to 1 \qquad \text{as} \quad n \to \infty$$

uniformly over every bounded interval. From this we obtain

$$\prod_{j=1}^{k_n} |\,f_{nj}(u)\,|^2 \to 1 \qquad \text{as} \quad n \to \infty$$

uniformly over every bounded interval. Over any bounded interval and for sufficiently large values of n we may take logarithms and thus assert that

$$\sum_{j=1}^{k_n} \log\,(1 - (1 - |\,f_{nj}(u)\,|^2)) \to 0 \qquad \text{as} \quad n \to \infty$$

uniformly over every bounded interval. Using the easily verified inequality $-\log\,(1 - x) \geq x$ for $0 \leq x < 1$, we obtain for sufficiently large n,

$$-\sum_{j=1}^{k_n} \log\,(1 - (1 - |\,f_{nj}(u)\,|^2)) \geq \sum_{j=1}^{k_n} (1 - |\,f_{nj}(u)\,|^2) \geq 0,$$

or $\sum_{j=1}^{k_n}(1 - |\,f_{nj}(u)\,|^2) \to 0$ as $n \to \infty$ uniformly over every bounded interval. Thus

$$(3) \qquad \int_{-1}^{+1} \{\sum_{j=1}^{k_n} (1 - |\,f_{nj}(u)\,|^2)\}\,du \to 0 \qquad \text{as} \quad n \to \infty.$$

Let $\mathcal{Y} = \{\{Y_{n,1}, \cdots, Y_{n,k_n}\}\}$ be random variables such that, for every n, the random variables $X_{n,1},\ Y_{n,1},\ \cdots,\ X_{n,k_n},\ Y_{n,k_n}$ are independent and such that F_{nj} is the distribution function of Y_{nj}, that is, $\mathcal{L}(X_{nj}) = \mathcal{L}(Y_{nj})$. If we let G_{nj} denote the distribution function of $X_{nj} - Y_{nj}$, then $|\,f_{nj}(u)\,|^2$ is the characteristic function of G_{nj}. By Fubini's theorem and easy integration we obtain

$$\int_{-1}^{+1} (1 - |f_{nj}(u)|^2)\, du = 2 \int_{-\infty}^{\infty} \left(1 - \frac{\sin x}{x}\right) dG_{nj}(x).$$

We note that for $|x| > 1$, $1 - (\sin x)/x \geq K_1 > 0$, where K_1 is some positive constant, and for $|x| \leq 1$ we have

$$1 - \frac{\sin x}{x} = 1 - \frac{1}{x}\left(x - \frac{x^3}{3!} + \frac{x^5}{5!} - + \cdots\right) \geq K_2 x^2,$$

where $K_2 > 0$. Let $K_3 = \min\{K_1, K_2\} > 0$, and using Lemma 1 we obtain

$$\int_{-1}^{1} \left\{ \sum_{j=1}^{k_n} (1 - |f_{nj}(u)|^2) \right\} du$$

$$\geq K_3 \left\{ \sum_{j=1}^{k_n} \int_{|x| \leq 1} x^2\, dG_{nj}(x) + \sum_{j=1}^{k_n} \int_{|x| > 1} dG_{nj}(x) \right\}$$

$$= K_3 \sum_{j=1}^{k_n} E\psi(X_{nj} - Y_{nj})$$

$$\geq \frac{K_3}{2} \sum_{j=1}^{k_n} E\psi(X_{nj} - m_{nj})$$

$$= \frac{K_3}{2} \left\{ \sum_{j=1}^{k_n} \int_{|x| \leq 1} x^2\, dF_{nj}(x + m_{nj}) \right.$$

$$\left. + \sum_{j=1}^{k_n} \int_{|x| > 1} dF_{nj}(x + m_{nj}) \right\} \geq 0.$$

By (3) this last expression converges to zero. Since each of the two sums in this last expression are nonnegative, they each converge to zero, proving that (a) and (b) are necessary. Q.E.D.

Corollary to Theorem 1. Let $\mathfrak{X} = \{\{X_{n,1}, \cdots, X_{n,k_n}\}\}$ be a double sequence of random variables which are row-wise independent. There exists a sequence of constants $\{c_n\}$ such that $X_{n,1} + \cdots + X_{n,k_n} - c_n \xrightarrow{P} 0$

if and only if

$$\sum_{j=1}^{k_n} E\left\{\frac{(X_{nj} - m_{nj})^2}{1 + (X_{nj} - m_{nj})^2}\right\} \to 0 \quad \text{as} \quad n \to \infty.$$

Proof: It is easy to verify that the function ψ defined in Lemma 1 satisfies $x^2/(1 + x^2) \leq \psi(x) \leq 2x^2/(1 + x^2)$. The corollary then immediately follows from Theorem 1.

Theorem 2. Let $\mathfrak{X} = \{\{X_{n,1}, \cdots, X_{n,k_n}\}\}$ be a double sequence of random variables which are row-wise independent. If $\{c_n\}$ is a sequence of constants such that the distribution function of $X_{n,1} + \cdots + X_{n,k_n} - c_n$ converges completely to some distribution function, then there is a constant $C > 0$ such that for all n

$$\sum_{j=1}^{k_n} \int_{-\infty}^{\infty} \frac{x^2}{1 + x^2} \, dF_{nj}(x + m_{nj}) < C.$$

Proof: Let $Z_n = X_{n,1} + \cdots + X_{n,k_n} - c_n$. We are given that $\mathcal{L}(Z_n) \to \mathcal{L}(Z)$ as $n \to \infty$ for some random variable Z. Let a_n be any sequence of real numbers such that $a_n \to 0$ as $n \to \infty$. By Theorem 3 in Section 4.2, $\mathcal{L}(a_n Z_n) \to \mathcal{L}(0)$ or $a_n Z_n \xrightarrow{P} 0$. Let $0 < c < 1$. Then for every random variable Y,

$$E\left\{\frac{c^2 Y^2}{1 + c^2 Y^2}\right\} \geq c^2 E\left\{\frac{Y^2}{1 + Y^2}\right\}.$$

Let us denote

$$S_n = \sum_{j=1}^{k_n} E\left\{\frac{a_n^2(X_{nj} - m_{nj})^2}{1 + a_n^2(X_{nj} - m_{nj})^2}\right\}.$$

Then

$$S_n \geq a_n^2 \sum_{j=1}^{k_n} E\left\{\frac{(X_{nj} - m_{nj})^2}{1 + (X_{nj} - m_{nj})^2}\right\}.$$

By the corollary to Theorem 1, $a_n Z_n \xrightarrow{P} 0$ implies that $S_n \to 0$ as $n \to \infty$. Thus, since the right side of the last inequality above converges to zero as $n \to \infty$ for *all* sequences $\{a_n\}$ for which $a_n \to 0$ as $n \to \infty$, it follows that the sequence of sums

$$\sum_{j=1}^{k_n} E\left\{\frac{(X_{nj} - m_{nj})^2}{1 + (X_{nj} - m_{nj})^2}\right\}, \qquad n = 1, 2, \cdots,$$

is bounded. Q.E.D.

Lemma 2. Let $\mathfrak{X} = \{\{X_{n,1}, \cdots, X_{n,k_n}\}\}$ be an infinitesimal system. Then

(a) $\qquad \max_{1 \leq j \leq k_n} |m_{nj}| \to 0 \qquad$ as $\quad n \to \infty$, and

(b) $\qquad \max_{1 \leq j \leq k_n} |a_{nj}| \to 0 \qquad$ as $\quad n \to \infty$,

where

$$a_{nj} = \int_{|x| < \tau} x \, dF_{nj}(x) \quad \text{and} \quad \tau > 0.$$

Proof: Let $\{j_n\}$ be a sequence of positive integers which satisfy $1 \leq j_n \leq k_n$ and $|m_{nj_n}| = \max_{1 \leq j \leq k_n} |m_{nj}|$. Since $\mathfrak{X}$ is assumed to be an infinitesimal system, then $X_{nj_n} \xrightarrow{P} 0$. Let $\epsilon > 0$ be arbitrary. Then there is an N such that $n > N$ implies $P[\,|X_{nj_n}| < \epsilon\,] > \frac{3}{4}$. This implies that $\limsup_{n \to \infty} |m_{nj_n}| \leq \epsilon$, and the arbitrariness of $\epsilon > 0$ implies that $|m_{nj_n}| = \max_{1 \leq j \leq k_n} |m_{nj}| \to 0$ as $n \to \infty$. Next let $\{i_n\}$ be defined by $1 \leq i_n \leq k_n$ and $|a_{ni_n}| = \max_{1 \leq j \leq k_n} |a_{nj}|$. Let $\epsilon > 0$ be arbitrarily small, $0 < \epsilon < \tau$. We obtain

$$|a_{ni_n}| \leq \left| \int_{|x| < \epsilon} x \, dF_{ni_n}(x) \right| + \left| \int_{\epsilon \leq |x| < \tau} x \, dF_{ni_n}(x) \right|$$

$$\leq \epsilon + \tau P[\,|X_{ni_n}| \geq \epsilon\,].$$

Hence $\limsup_{n \to \infty} |a_{ni_n}| \leq \epsilon$ for ϵ arbitrary, and so (b) is proved. Q.E.D.

The first two theorems and the first lemma of this section were not concerned with infinitesimal systems but only with double sequences of random variables which are row-wise independent. However, they are actually lemmas for the really important Theorem 3 which *is* a theorem about infinitesimal systems.

Theorem 3. If $\mathfrak{X} = \{\{X_{n,1}, \cdots, X_{n,k_n}\}\}$ is an infinitesimal system, if $\{c_n\}$ is a sequence of constants, and if the distribution function of $X_{n,1} + \cdots + X_{n,k_n} - c_n$ converges completely, then there is a $C > 0$ such that

$$\sum_{j=1}^{k_n} \int \frac{x^2}{1 + x^2} dF_{nj}(x + a_{nj}) < C$$

for all n, where

$$a_{nj} = \int_{|x| < \tau} x \, dF_{nj}(x) \quad \text{and} \quad \tau > 0.$$

Proof: Let us denote

$$S_n = \sum_{j=1}^{k_n} \int \frac{x^2}{1 + x^2} dF_{nj}(x + a_{nj}).$$

In proving $S_n < C$ we shall twice use the easily verified inequality: $(a + b)^2 \leq 2(a^2 + b^2)$. We may write

$$S_n = \sum_{j=1}^{k_n} \int \frac{(x + m_{nj} - a_{nj})^2}{1 + (x + m_{nj} - a_{nj})^2} dF_{nj}(x + m_{nj}).$$

Thus $S_n \leq K_n + L_n$, where

$$K_n = 2 \sum_{j=1}^{k_n} \int \frac{x^2}{1 + (x + m_{nj} - a_{nj})^2} dF_{nj}(x + m_{nj})$$

and $L_n = 2 \sum_{j=1}^{k_n} (m_{nj} - a_{nj})^2$. Let $\epsilon > 0$ be arbitrary. By Lemma 2 we may assert that there is a constant $K_0 > 0$ such that

$$\frac{x^2}{1 + (x + m_{nj} - a_{nj})^2} \leq K_0 \frac{x^2}{1 + x^2}$$

for all $|x| \geq \epsilon$. Also

$$\int_{|x| < \epsilon} \frac{x^2}{1 + (x + m_{nj} - a_{nj})^2} dF_{nj}(x + m_{nj})$$

$$\leq \int_{|x| < \epsilon} x^2 \, dF_{nj}(x + m_{nj})$$

$$\leq (1 + \epsilon^2) \int_{|x| < \epsilon} \frac{x^2}{1 + x^2} dF_{nj}(x + m_{nj}).$$

Let $K = \max\{K_o, (1 + \epsilon^2)\}$. We obtain then that

$$K_n \leq 2K \sum_{j=1}^{k_n} \int \frac{x^2}{1 + x^2}\, dF_{nj}(x + m_{nj})$$

for all n. We next find an upper bound for L_n. We first observe that

$$(a_{nj} - m_{nj})^2 = \left(\int_{|x| < \tau} (x - m_{nj})\, dF_{nj}(x) - \int_{|x| \geq \tau} m_{nj}\, dF_{nj}(x) \right)^2$$

$$\leq 2 \left(\int_{|x + m_{nj}| < \tau} |x|\, dF_{nj}(x + m_{nj}) \right)^2$$

$$+ 2 \left(\int_{|x + m_{nj}| \geq \tau} |m_{nj}|\, dF_{nj}(x + m_{nj}) \right)^2.$$

Since by Lemma 2, $\max_{1 \leq j \leq k_n} |m_{nj}| \to 0$ as $n \to \infty$, we have, for all sufficiently large n, that

$$(a_{nj} - m_{nj})^2 \leq 2 \left(\int_{|x| < 2\tau} |x|\, dF_{nj}(x + m_{nj}) \right)^2$$

$$+ 2m_{nj}^2 \int_{|x| > \tau/2} dF_{nj}(x + m_{nj}).$$

By Schwartz's inequality we obtain

$$\sum_{j=1}^{k_n} (a_{nj} - m_{nj})^2 \leq 2 \sum_{j=1}^{k_n} \int_{|x| < 2\tau} x^2\, dF_{nj}(x + m_{nj})$$

$$+ 2 \left(\max_{1 \leq j \leq k_n} |m_{nj}|^2 \right) \sum_{j=1}^{k_n} \int_{|x| > \tau/2} dF_{nj}(x + m_{nj}).$$

Hence there exists a constant M such that $L_n \leq MT_n$ for all n, where

$$T_n = \sum_{j=1}^{k_n} \int \frac{x^2}{1 + x^2}\, dF_{nj}(x + m_{nj}),$$

and hence $S_n \leq (2K + M)T_n$ for all n. But by the hypothesis and by Theorem 2, $\{T_n\}$ is bounded, and thus we obtain the conclusion of the theorem.

EXERCISES

1. Prove: If X and Y are independent identically distributed random variables with common characteristic function f, then the characteristic function of $X - Y$ is $|f(u)|^2$.

2. Prove: If $\{X_n\}$ is a sequence of random variables such that $\mathcal{L}(X_n) \to \mathcal{L}(0)$ as $n \to \infty$, then $X_n \overset{P}{\to} 0$.

3. Prove: If $\{X_n\}$ is a sequence of random variables such that $\mathcal{L}(X_n) \to \mathcal{L}(X)$ as $n \to \infty$, and if $\{a_n\}$ is a sequence of real numbers such that $a_n \to a$ as $n \to \infty$, then $\mathcal{L}(a_n X_n) \to \mathcal{L}(aX)$ as $n \to \infty$.

4. Let $\{X_n\}$ be a sequence of independent identically distributed random variables, let $k_n = n$, and define $X_{nj} = X_j/\sqrt{n}$, $1 \le j \le k_n = n$, $n = 1, 2, \cdots$. Prove that $\mathfrak{X} = \{\{X_{n,1}, \cdots, X_{n,k_n}\}\}$ is an infinitesimal system.

5. Let X, Y be random variables, and let $B = [X = Y]$. Verify that $P([\,|X| > K] \cap B) = P([\,|Y| > K] \cap B) \le P[\,|Y| > K]$.

6. Prove: if $P[\,|X| < \epsilon] > \frac{2}{3}$, then every median of X is in $[-\epsilon, \epsilon]$.

6.5. The General Limit Theorem for Sequences of Sums of Independent Random Variables

The purpose of this section is to obtain Theorem 3 which gives necessary and sufficient conditions that the distribution functions of a sequence of centered sums of independent random variables from an infinitesimal system converge completely. The limit distribution function is an infinitely divisible distribution function, and it turns out that all infinitely divisible distribution functions can be obtained as such limiting distributions.

Lemma 1. For all real θ the following three inequalities hold:

(a) $|e^{i\theta} - 1| \le |\theta|$,

(b) $|e^{i\theta} - 1 - i\theta| \le 2\,|\theta|$, and

(c) $|e^{i\theta} - 1 - i\theta| \le \theta^2/2$.

Proof: We first observe that

$$i\int_0^\theta e^{it}\,dt = e^{i\theta} - 1.$$

But

$$\left| \int_0^\theta e^{it}\,dt \right| \le \int_0^{|\theta|} |e^{it}|\,dt \le |\theta|,$$

and thus we obtain (a). Inequality (b) is a trivial consequence of (a). In order to prove (c), we note that

$$i \int_0^\theta (e^{it} - 1)\,dt = e^{i\theta} - 1 - i\theta.$$

However, by (a),

$$\left| i \int_0^\theta (e^{it} - 1)\,dt \right| \le \int_0^{|\theta|} t\,dt = \tfrac{1}{2}\theta^2,$$

from which we obtain (c). Q.E.D.

Throughout this section, $\mathfrak{X} = \{\{X_{nj}, 1 \le j \le k_n\}\}$ will denote an infinitesimal system of random variables as defined in Section 6.4. Recall that for each n, $X_{n,1}, \cdots, X_{n,k_n}$ are independent. The sequence $\{Z_n\}$ defined by $Z_n = X_{n1} + \cdots + X_{nk_n} - c_n$ will be referred to as a *sequence of centered sums of independent random variables from the infinitesimal system* $\mathfrak{X}$. The constants $\{c_n\}$ will be called the *centering constants*. If $F_{Z_n} \xrightarrow{c} F$, we shall refer to the distribution function F (or to its characteristic function) as the *limit law* of the sequence $\{Z_n\}$ or of the sequence $\{F_{Z_n}\}$. We continue in our use of the notation:

$$F_{nj}(x) = P[X_{nj} \le x], \qquad f_{nj}(u) = E \exp (iuX_{nj}),$$

and

$$a_{nj} = \int_{|x|<\tau} x\,dF_{nj}(x).$$

Theorem 1. Let $\mathfrak{X} = \{\{X_{nj}\}\}$ be an infinitesimal system and let $\{\{a_{nj}\}\}$, $\{\{F_{nj}\}\}$, and $\{Z_n\}$ be as defined above. A necessary and sufficient condition that $\{F_{Z_n}\}$ converge completely to some limit law F for a sequence of centering constants $\{c_n\}$ is that the characteristic functions

$$(1) \quad \varphi_n(u) = \exp \left\{ -ic_n u \right.$$

$$\left. + \sum_{j=1}^{k_n} \left\{ iua_{nj} + \int_{-\infty}^{\infty} (e^{iux} - 1)\,dF_{nj}(x + a_{nj}) \right\} \right\}$$

converge (as $n \to \infty$) to some characteristic function φ. In either case, φ is the characteristic function of F.

Proof: It should be pointed out first of all that $\varphi_n(u)$ as defined in (1) is indeed a characteristic function; this follows from Theorem 6 in Section 6.1. By the continuity theorem a necessary and sufficient condition that $F_{Z_n} \xrightarrow{c} F$ is that

$$(2) \qquad f_n(u) = \exp(-ic_nu) \prod_{j=1}^{k_n} f_{nj}(u) \to f(u)$$

as $n \to \infty$, where $f(u)$ is the characteristic function of F. Let us denote

$$f'_{nj}(u) = \int_{-\infty}^{\infty} e^{iux}\, dF_{nj}(x + a_{nj}) = E \exp iu(X_{nj} - a_{nj}).$$

Then (2) may be rewritten

$$f_n(u) = \{\exp[-iu(c_n - \sum_{j=1}^{k_n} a_{nj})]\} \prod_{j=1}^{k_n} f'_{nj}(u) \to f(u)$$

as $n \to \infty$. Let us denote $\beta_{nj}(u) = f'_{nj}(u) - 1$. By Proposition 2 and Lemma 2 in Section 6.4 we have $\max_{1 \le j \le k_n} |\beta_{nj}(u)| \to 0$ as $n \to \infty$ uniformly over every bounded set. We consider an arbitrary (fixed!) u. Hence for sufficiently large n (and here the logarithm as defined in Section 4.3 and its principal value are the same),

$$\log f'_{nj}(u) = \log(1 + \beta_{nj}(u))$$

$$= \beta_{nj}(u) - \frac{\beta_{nj}^2(u)}{2} + \frac{\beta_{nj}^3(u)}{3} - + \cdots,$$

and thus for sufficiently large n

$$(3) \quad |\sum_{j=1}^{k_n} \log f'_{nj}(u) - \sum_{j=1}^{k_n} \beta_{nj}(u)| \le \sum_{j=1}^{k_n} \sum_{s=2}^{\infty} \frac{|\beta_{nj}(u)|^s}{s}$$

$$\le \frac{1}{2} \sum_{j=1}^{k_n} \frac{|\beta_{nj}(u)|^2}{1 - |\beta_{nj}(u)|}$$

$$\le (\max_{1 \le j \le k_n} |\beta_{nj}(u)|) \sum_{j=1}^{k_n} |\beta_{nj}(u)|.$$

We first find an upper bound for the sum $\sum_{j=1}^{k_n} |\beta_{nj}(u)|$ before we turn toward the proof of either the necessity or the sufficiency of the condition. We observe that

$$|\beta_{nj}(u)| = \left| \int_{|x|<\tau} (e^{iux} - 1 - iux) \, dF'_{nj}(x) \right.$$

$$+ \int_{|x|\geq\tau} (e^{iux} - 1) \, dF'_{nj}(x)$$

$$+ iu \int_{|x|<\tau} x \, dF'_{nj}(x) \Bigg|.$$

By Lemma 1 we obtain

$$(4) \qquad |\beta_{nj}(u)| \leq \frac{u^2}{2} \int_{|x|<\tau} x^2 \, dF'_{nj}(x) + 2 \int_{|x|\geq\tau} dF'_{nj}(x)$$

$$+ \left| u \int_{|x|<\tau} x \, dF'_{nj}(x) \right|.$$

We now find a bound for the third term on the right side of (4). By Lemma 2 in Section 6.4 we have $|a_{nj}| < \tau/2$ for large n. Thus,

$$(5) \qquad \left| \int_{|x|<\tau} x \, dF'_{nj}(x) - \int_{|x+a_{nj}|<\tau} x \, dF'_{nj}(x) \right|$$

$$\leq \int_{\tau/2<|x|<3\tau/2} |x| \, dF'_{nj}(x)$$

$$\leq \frac{3\tau}{2} \int_{|x|>\tau/2} dF'_{nj}(x).$$

Also, for large n,

$$(6) \qquad \left| \int_{|x+a_{nj}|<\tau} x \, dF'_{nj}(x) \right| = \left| \int_{|x|<\tau} (x - a_{nj}) \, dF_{nj}(x) \right|$$

$$= \left| a_{nj} \left(1 - \int_{|x|<\tau} dF_{nj}(x) \right) \right|$$

$$= |a_{nj}| \int_{|x|\geq\tau} dF_{nj}(x) \leq \frac{\tau}{2} \int_{|x|>\tau/2} dF'_{nj}(x).$$

Combining (5) and (6) we obtain

(7)
$$\left| \int_{|x|<\tau} x \, dF'_{nj}(x) \right| \leq 2\tau \int_{|x|>\tau/2} dF'_{nj}(x).$$

Thus, by (4) and (7) we have

(8)
$$\sum_{j=1}^{k_n} |\beta_{nj}(u)| \leq \frac{u^2}{2} \sum_{j=1}^{k_n} \int_{|x|<\tau} x^2 \, dF'_{nj}(x)$$

$$+ (2 + 2\tau |u|) \sum_{j=1}^{k_n} \int_{|x|>\tau/2} dF'_{nj}(x).$$

We now prove that the condition is necessary. We are given $F_{Z_n} \xrightarrow{c} F$. We denote

$$S_n = \sum_{j=1}^{k_n} \int_{-\infty}^{\infty} \frac{x^2}{1+x^2} dF'_{nj}(x).$$

Hence, by Theorem 3 in Section 6.4, for some $C > 0$,

(9)
$$S_n \leq C \quad \text{for all} \quad n.$$

Thus,

(10)
$$\sum_{j=1}^{k_n} \int_{|x|<\tau} x^2 \, dF'_{nj}(x) \leq (1 + \tau^2) S_n \leq (1 + \tau^2) C$$

for all n. Also, for all n,

(11)
$$\sum_{j=1}^{k_n} \int_{|x|>\tau/2} dF'_{nj}(x) \leq \left(\frac{1 + (\tau/2)^2}{(\tau/2)^2} \right) S_n$$

$$\leq \frac{C(1 + (\tau/2)^2)}{(\tau/2)^2}.$$

By (8), (10) and (11) we obtain that $\sum_{j=1}^{k_n} |\beta_{nj}(u)|$ is bounded by a constant which does not depend on u or n. Hence, by (3) and the already proved fact that $\max_{1 \leq j \leq k_n} |\beta_{nj}(u)| \to 0$ as $n \to \infty$ we obtain that

$$\left| \sum_{j=1}^{k_n} \log f'_{nj}(u) - \sum_{j=1}^{k_n} \beta_{nj}(u) \right| \to 0 \qquad \text{as} \quad n \to \infty.$$

Finally, using the definition of logarithm as given in Section 4.3,

$$(12) \quad |\log f_n(u) - \log \varphi_n(u)|$$

$$= \left| -i\left(c_n - \sum_{j=1}^{k_n} a_{nj}\right)u + \sum_{j=1}^{k_n} \log f'_{nj}(u) \right.$$

$$+ i\left(c_n - \sum_{j=1}^{k_n} a_{nj}\right)u - \sum_{j=1}^{k_n} \beta_{nj}(u) \Big|$$

$$= \left| \sum_{j=1}^{k_n} \log f'_{nj}(u) - \sum_{j=1}^{k_n} \beta_{nj}(u) \right| \to 0 \qquad \text{as} \quad n \to \infty.$$

Thus, if $F_{Z_n} \xrightarrow{c} F$, then $\varphi_n(u) \to f(u)$ as $n \to \infty$ for all u.

We now prove that the condition is sufficient. Let us denote

$$G_n(x) = \sum_{j=1}^{k_n} \int_{-\infty}^{x} \frac{t^2}{1 + t^2} \, dF'_{nj}(t).$$

Then we may rewrite $\varphi_n(u)$ as

$$\varphi_n(u) = \exp\left\{ i\gamma_n u + \int_{-\infty}^{\infty} \left(e^{iux} - 1 - \frac{iux}{1 + x^2} \right) \frac{1 + x^2}{x^2} \, dG_n(x) \right\},$$

where

$$\gamma_n = -c_n + \sum_{j=1}^{k_n} \left(a_{nj} + \int_{-\infty}^{\infty} \frac{\tau}{1 + \tau^2} \, dF'_{nj}(\tau) \right).$$

The functions $\{\varphi_n(u)\}$ are characteristic functions, and by hypothesis $\varphi_n(u) \to \varphi(u)$ as $n \to \infty$, where $\varphi(u)$ is a characteristic function. By Theorem 5 of Section 6.1, $\varphi(u)$ is an infinitely divisible characteristic function determined by some pair (γ, G). By Theorem 1 of Section 6.3, $\gamma_n \to \gamma$ and $G_n \xrightarrow{c} G$ as $n \to \infty$. In particular, $G_n(+\infty) = S_n \to G(+\infty)$ as $n \to \infty$, since $G_n \xrightarrow{c} G$. Hence $\{S_n\}$ are bounded. Thus (9) holds if $\varphi_n \to \varphi$ as $n \to \infty$. Since (10) and (11) hold, (12) follows, from which we obtain that $f_n(u) \to \varphi(u) = f(u)$ as $n \to \infty$. Q.E.D.

As a corollary to Theorem 1 we obtain the following theorem due to A. Ya. Khinchine.

Theorem 2. A necessary and sufficient condition that a distribution function F be the limit law of a sequence of centered sums of independent random variables from an infinitesimal system is that F be an infinitely divisible distribution function.

Proof: The necessity of the condition follows immediately from Theorem 5 in Section 6.1 and Theorem 1 above. Conversely, suppose F is an infinitely divisible distribution function with characteristic function f. Let $\{X_{n,j}, 1 \leq j \leq n, n = 1, 2, \cdots\}$ all be independent random variables where the characteristic function of X_{nj} is $\exp\{(1/n) \log f(u)\}$. Then it is easy to verify that $\{\{X_{nj}\}\}$ is an infinitesimal system. Since $F_{Z_n} = F$, where $Z_n = X_{n,1} + \cdots + X_{n,n}$, the conclusion is established.

The following elementary lemma will be needed in the proof of Theorem 3.

Lemma 2. If φ, φ_1, φ_2, $\cdots$ are nondecreasing functions over an interval I, if $\varphi_n(x) \to \varphi(x)$ as $n \to \infty$ at all $x \in Cont\ \varphi$, if $x_n \in I$ and $x_n \to x_0 \in Cont\ \varphi$, then $\varphi_n(x_n) \to \varphi(x_0)$ as $n \to \infty$.

Proof: Let $\epsilon > 0$ be arbitrary. Then there exists a $\delta > 0$ such that $|\varphi(x_0) - \varphi(y)| < \epsilon/2$ for all y satisfying $|x_0 - y| < \delta$. Let $\{y_1, y_2\} \subset Cont\ \varphi$ satisfy

$$x_0 - \delta < y_1 < x_0 < y_2 < x_0 + \delta.$$

Then there is a positive integer N such that

$$\max\{|\varphi_n(y_1) - \varphi(y_1)|, |\varphi_n(x_0) - \varphi(x_0)|, |\varphi_n(y_2) - \varphi(y_2)|\} < \epsilon/2$$

and $\min\{y_2 - x_0, x_0 - y_1\} > |x_0 - x_n|$ for all $n > N$. Hence, because φ_n, φ are nondecreasing, we have for $n > N$

$$\varphi_n(x_n) \geq \varphi_n(y_1) > \varphi(y_1) - \epsilon/2 > \varphi(x_0) - \epsilon,$$

and

$$\varphi_n(x_n) \leq \varphi_n(y_2) < \varphi(y_2) + \epsilon/2 < \varphi(x_0) + \epsilon,$$

which concludes the proof.

Definition: A real-valued function M is called a Lévy spectral function if and only if

   (a) M is defined over $(-\infty, 0) \cup (0, \infty)$,

   (b) M is nondecreasing over $(-\infty, 0)$ and over $(0, \infty)$,

   (c) $M(-\infty) = M(+\infty) = 0$, and

   (d) $\displaystyle\int_{-1}^{-0} + \int_{+0}^{+1} x^2 \, dM(x) < \infty.$

Our most general limit theorem is the following:

Theorem 3. Let $\mathfrak{X} = \{\{X_{n,1}, \cdots, X_{n,k_n}\}\}$ be an infinitesimal system of random variables, let $\{c_n\}$ be a sequence of real numbers, and let $Z_n = X_{n1} + \cdots + X_{nk_n} - c_n$. In order that there exist a distribution function F such that $F_{Z_n} \xrightarrow{c} F$ it is necessary and sufficient that there exist constants γ and $\sigma^2 \geq 0$ and a Lévy spectral function M such that

$$(13) \quad M(x) = \begin{cases} \displaystyle\lim_{n \to \infty} \sum_{j=1}^{k_n} F_{nj}(x) & \text{if } 0 > x \in \text{Cont } M \\[2em] \displaystyle\lim_{n \to \infty} \sum_{j=1}^{k_n} (F_{nj}(x) - 1) & \text{if } 0 < x \in \text{Cont } M, \end{cases}$$

$$(14) \quad \lim_{\epsilon \downarrow 0} \begin{Bmatrix} \text{lim sup} \\ \text{lim inf} \end{Bmatrix}_{n \to \infty} \sum_{j=1}^{k_n} \left\{ \int_{|x|<\epsilon} x^2 \, dF_{nj}(x) - \left(\int_{|x|<\epsilon} x \, dF_{nj}(x) \right)^2 \right\} = \sigma^2,$$

and

$$(15) \quad \sum_{j=1}^{k_n} a_{nj} - c_n \to \gamma + \int_{|x|<\tau} \frac{x^3}{1+x^2} \, dM(x) - \int_{|x|>\tau} \frac{x}{1+x^2} \, dM(x)$$

as $n \to \infty$, where

$$a_{nj} = \int_{|x|<\tau} x \, dF_{nj}(x), \qquad \tau > 0,$$

and $\pm\tau \in \text{Cont } M$. The distribution function F is infinitely divisible and $\{\gamma, \sigma^2, M\}$ determine F as in Theorem 2 in Section 6.2.

Proof: Select $\tau > 0$; at an opportune subsequent moment it can be shifted slightly in order that $\pm\tau \in \mathit{Cont}\ M$. Then define

$$a_{nj} = \int_{|x|<\tau} x\, dF_{nj}(x),$$

$$M_n(x) = \begin{cases} \sum_{j=1}^{k_n} F_{nj}(x + a_{nj}) & \text{if}\quad x < 0 \\[2em] \sum_{j=1}^{k_n} (F_{nj}(x + a_{nj}) - 1) & \text{if}\quad x > 0, \end{cases}$$

$\sigma_n^2 = 0$, and

$$\gamma_n(\tau) = -c_n + \sum_{j=1}^{k_n} a_{nj} + \sum_{j=1}^{k_n} \int_{-\infty}^{\infty} \frac{x}{1 + x^2}\, dF_{nj}(x + a_{nj}).$$

By Theorems 1 and 2 a necessary and sufficient condition that $F_{Z_n} \xrightarrow{c} F$ is that the characteristic functions

$$\varphi_n(u) = \exp\left\{ -ic_n u + \sum_{j=1}^{k_n} \left\{ iua_{nj} + \int_{-\infty}^{\infty} (e^{iux} - 1)\, dF_{nj}(x + a_{nj}) \right\} \right\}$$

converge to the characteristic function φ of F, where φ is necessarily infinitely divisible and determined by a triple $\{\gamma, \sigma^2, M\}$. We may rewrite φ_n as

$$\varphi_n(u) = \exp\left\{ i\gamma_n(\tau)u + \int_{-\infty}^{\infty} \left(e^{iux} - 1 - \frac{iux}{1 + x^2} \right) dM_n(x) \right\}.$$

By Theorem 2 of Section 6.3, a necessary and sufficient condition that $\varphi_n(u) \to \varphi(u)$ is that

$$(13') \qquad M_n(x) \to M(x) \qquad \text{at all}\quad x \in \mathit{Cont}\ M,$$

$$(14') \quad \lim_{\epsilon \downarrow 0} \begin{Bmatrix} \lim\sup \\ \lim\inf \end{Bmatrix}_{n \to \infty} \left\{ \int_{-\epsilon}^{\epsilon} x^2\, dM_n(x) + \sigma_n^2 \right\}$$

$$= \lim_{\epsilon \downarrow 0} \begin{Bmatrix} \lim\sup \\ \lim\inf \end{Bmatrix}_{n \to \infty} \sum_{j=1}^{k_n} \int_{-\epsilon}^{\epsilon} x^2\, dF_{nj}(x + a_{nj}) = \sigma^2,$$

and

$$(15') \qquad \gamma_n(\tau) \to \gamma \qquad \text{as} \quad n \to \infty.$$

In order to prove the theorem we must show that under the general hypothesis of $\mathfrak{X}$ being an infinitesimal system, then $\{(13), (14), (15)\}$ and $\{(13'), (14'), (15')\}$ are equivalent. Whichever set of hypotheses is assumed true, we have in our possession a Lévy spectral function M and may assume $\tau > 0$ is such that $\{\tau, -\tau\} \subset Cont\, M$. We prove this theorem in three stages.

Stage (i): *Proof that* (13) *and* (13') *are equivalent.* Let $x < 0$ be such that $x \in Cont\, M$. Let $a_n = \max\{\,|a_{nj}|:1 \le j \le k_n\}$. By Lemma 2 in Section 6.4, $a_n \to 0$ as $n \to \infty$. Since

$$\sum_{j=1}^{k_n} \int_{-\infty}^{x-a_n} dF_{nj}(t + a_{nj}) \le \sum_{j=1}^{k_n} \int_{-\infty}^{x} dF_{nj}(t)$$

$$\le \sum_{j=1}^{k_n} \int_{-\infty}^{x+a_n} dF_{nj}(t + a_{nj})$$

$$\le \sum_{j=1}^{k_n} \int_{-\infty}^{x+2a_n} dF_{nj}(t),$$

we may conclude, by Lemma 2, that (13) and (13') are equivalent for $0 > x \in Cont\, M$. The same proof holds for $0 < x \in Cont\, M$.

Stage (ii): *Proof that* $\{(13), (14)\}$ *and* $\{(13'), (14')\}$ *are equivalent.* We first observe that

$$(16) \quad \sum_{j=1}^{k_n} \int_{|x|<\epsilon} (x - a_{nj})^2 \, dF_{nj}(x)$$

$$= \sum_{j=1}^{k_n} \left\{ \int_{|x|<\epsilon} x^2 \, dF_{nj}(x) - 2a_{nj} \int_{|x|<\epsilon} x \, dF_{nj}(x) + a_{nj}^2 \right.$$

$$\left. - a_{nj}^2 \int_{|x|\ge\epsilon} dF_{nj}(x) \right\}$$

$$= \sum_{j=1}^{k_n} \left\{ \int_{|x|<\epsilon} x^2 \, dF_{nj}(x) - \left(\int_{|x|<\epsilon} x \, dF_{nj}(x) \right)^2 \right\} + W_n,$$

where

$$W_n = \sum_{j=1}^{k_n} \left\{ \left(\int_{|x|<\epsilon} x \, dF_{nj}(x) \right)^2 - 2a_{nj} \int_{|x|<\epsilon} x \, dF_{nj}(x) \right.$$

$$\left. + a^2_{nj} - a^2_{nj} \int_{|x|\geq\epsilon} dF_{nj}(x) \right\}.$$

We now show that when (13) and (13′) are true, then $W_n \to 0$ as $n \to \infty$. Clearly, for $0 < \epsilon < \tau$,

$$W_n = \sum_{j=1}^{k_n} \left\{ \left(\int_{\epsilon \leq |x| < \tau} x \, dF_{nj}(x) \right)^2 - a^2_{nj} \int_{|x|\geq\epsilon} dF_{nj}(x) \right\}.$$

Thus,

$$|W_n| \leq \sum_{j=1}^{k_n} \left\{ \tau^2 \int_{\epsilon \leq |x| < \tau} dF_{nj}(x) + a^2_{nj} \right\} \int_{|x|\geq\epsilon} dF_{nj}(x)$$

$$\leq \left\{ \tau^2 \left(\max_{1 \leq j \leq k_n} \int_{|x|\geq\epsilon} dF_{nj}(x) \right) + a_n^2 \right\} \sum_{j=1}^{k_n} \int_{|x|\geq\epsilon} dF_{nj}(x).$$

Since (13) and (13′) have already been proved to be equivalent, whichever of $\{(13), (14)\}$, $\{(13'), (14')\}$ we assume to be true, (13) is nevertheless true, which implies that the sum on the right end above is bounded. Since $a_n \to 0$ and since $\mathfrak{X}$ is an infinitesimal system, the terms in the curly brackets above tend to zero, and consequently $W_n \to 0$ under (13) or (13′). Thus, by (16), (14) is true if and only if

$$\lim_{\epsilon \downarrow 0} \left. {\lim \sup \atop \lim \inf} \right\} \sum_{\substack{j=1 \\ n \to \infty}}^{k_n} \int_{|x|<\epsilon} (x - a_{nj})^2 \, dF_{nj}(x)$$

(17)

$$= \lim_{\epsilon \downarrow 0} \left. {\lim \sup \atop \lim \inf} \right\} \sum_{\substack{j=1 \\ n \to \infty}}^{k_n} \int_{|x+a_{nj}|<\epsilon} x^2 \, dF_{nj}(x + a_{nj}) = \sigma^2.$$

Since $a_n \to 0$, and since the quantity following $\lim_{\epsilon \downarrow 0}$ is monotone non-increasing in ϵ as $\epsilon \to 0$, we may conclude that (17) is true if and only if (14') is true. Thus if (13) or (13') is true, (14) is equivalent to (14').

Stage (iii): *Proof that* $\{(13), (14), (15)\}$ *and* $\{(13'), (14'), (15')\}$ *are equivalent.* Let us denote

$$L_n = \sum_{j=1}^{k_n} \int_{-\infty}^{\infty} \frac{x}{1 + x^2} \, dF_{nj}(x + a_{nj}).$$

We can prove (15) and (15') equivalent if we can prove under $\{(13), (14)\}$ or $\{(13'), (14')\}$ that

$$(18) \qquad L_n \to -\int_{|x| < \tau} \frac{x^3}{1 + x^2} \, dM(x) + \int_{|x| > \tau} \frac{x}{1 + x^2} \, dM(x)$$

as $n \to \infty$. It is easily observed that both integrals exist and are finite because of the properties of M. Let $T > \tau$, $T > 1$ and $\{+T, -T\} \subset Cont \, M$. Then we may write

$$L_n = \alpha_n(T) + \beta_n(T),$$

where

$$\alpha_n(T) = \sum_{j=1}^{k_n} \int_{|x| < T} \frac{x}{1 + x^2} \, dF_{nj}(x + a_{nj}),$$

and

$$\beta_n(T) = \sum_{j=1}^{k_n} \int_{|x| \geq T} \frac{x}{1 + x^2} \, dF_{nj}(x + a_{nj}).$$

We observe that (since $T > 1$)

$$|\beta_n(T)| \leq \frac{T}{1 + T^2} \sum_{j=1}^{k_n} \int_{|x| \geq T} dF_{nj}(x + a_{nj}).$$

Under either (13) or (13') this latter sum is bounded in n for fixed T and nonincreasing in T for fixed n. Hence, for T sufficiently large, $|\beta_n(T)|$ is arbitrarily small for all n. Next we notice that for sufficiently large n and $1 \leq j \leq k_n$,

$$\int_{|x| < T} \frac{x}{1 + x^2} \, dF_{nj}(x + a_{nj})$$

$$= \int_{|x - a_{nj}| < T} \frac{x - a_{nj}}{1 + (x - a_{nj})^2} \, dF_{nj}(x)$$

$$= \int_{|x - a_{nj}| < T} \frac{x}{1 + (x - a_{nj})^2} \, dF_{nj}(x)$$

$$- a_{nj} \int_{|x - a_{nj}| < T} \frac{1}{1 + (x - a_{nj})^2} \, dF_{nj}(x)$$

$$= \int_{|x| < \tau} x \, dF_{nj}(x) - a_{nj} \int_{-\infty}^{\infty} \frac{1}{1 + (x - a_{nj})^2} \, dF_{nj}(x)$$

$$+ a_{nj} \int_{|x - a_{nj}| \geq T} \frac{1}{1 + (x - a_{nj})^2} \, dF_{nj}(x) + M_{nj},$$

where

$$M_{nj} = - \int_{|x| < \tau} \frac{x(x - a_{nj})^2}{1 + (x - a_{nj})^2} \, dF_{nj}(x)$$

$$+ \int_{-T + a_{nj}}^{-\tau} + \int_{\tau}^{T + a_{nj}} \frac{x}{1 + (x - a_{nj})^2} \, dF_{nj}(x).$$

Hence,

$$\int_{|x| < T} \frac{x}{1 + x^2} \, dF_{nj}(x + a_{nj})$$

$$= a_{nj} \left\{ \int_{-\infty}^{\infty} \frac{(x - a_{nj})^2}{1 + (x - a_{nj})^2} \, dF_{nj}(x) \right.$$

$$\left. + \int_{|x - a_{nj}| \geq T} \frac{1}{1 + (x - a_{nj})^2} \, dF_{nj}(x) \right\} + M_{nj}.$$

Thus for sufficiently large n,

$$| \alpha_n(T) - \sum_{j=1}^{k_n} M_{nj} | \leq a_n \left\{ \sum_{j=1}^{k_n} \int_{-\infty}^{\infty} \frac{x^2}{1 + x^2} \, dF_{nj}(x + a_{nj}) \right.$$

$$\left. + \frac{1}{1 + T^2} \sum_{j=1}^{k_n} \int_{|x - a_{nj}| \geq T} dF_{nj}(x) \right\}.$$

By (13) or (13′), the last sum in the curly brackets is bounded. By $\{(13), (14)\}$ or $\{(13′), (14′)\}$, the first sum on the right is bounded. Since $a_n \to 0$, then

$$| \alpha_n(T) - \sum_{j=1}^{k_n} M_{nj} | \to 0 \qquad \text{as} \quad n \to \infty.$$

We finally turn our attention to $\sum_{j=1}^{k_n} M_{nj}$. We may write

$$\sum_{j=1}^{k_n} M_{nj} = A_n + B_n + C_n + D_n + E_n + F_n,$$

where

$$A_n = -\sum_{j=1}^{k_n} \int_{|x| < \tau} \frac{(x - a_{nj})^3}{1 + (x - a_{nj})^2} \, dF_{nj}(x),$$

$$B_n = -\sum_{j=1}^{k_n} a_{nj} \int_{|x| < \tau} \frac{(x - a_{nj})^2}{1 + (x - a_{nj})^2} \, dF_{nj}(x),$$

$$C_n = \sum_{j=1}^{k_n} \int_{|x| \geq \tau} \frac{x - a_{nj}}{1 + (x - a_{nj})^2} \, dF_{nj}(x),$$

$$D_n = \sum_{j=1}^{k_n} a_{nj} \int_{|x| \geq \tau} \frac{1}{1 + (x - a_{nj})^2} \, dF_{nj}(x),$$

$$E_n = -\sum_{j=1}^{k_n} \int_{|x - a_{nj}| \geq T} \frac{x - a_{nj}}{1 + (x - a_{nj})^2} \, dF_{nj}(x)$$

$$F_n = -\sum_{j=1}^{k_n} a_{nj} \int_{|x - a_{nj}| \geq T} \frac{1}{1 + (x - a_{nj})^2} \, dF_{nj}(x).$$

Let $\tau' > \tau$ be such that $\pm\tau' \in Cont\ M$ and $\tau' - \tau \geq a_n$ for all n. Also, denote

$$H_n(x) = \int_{-\tau'}^{x} \frac{t^3}{1 + t^2}\, dM_n(t), \qquad H(x) = \int_{-\tau'}^{x} \frac{t^3}{1 + t^2}\, dM(t).$$

We note that H is continuous at $x = 0$ and for $x \neq 0$, H and M share the same continuity points in $[-\tau', \tau']$. Hence by (13) or (13′), and (14′), $H_n \xrightarrow{c} H$. An easy computation yields

$$\left| A_n + \int_{-\tau}^{\tau} dH_n(x) \right| \leq \int_{-\tau-a_n}^{-\tau+a_n} + \int_{\tau-a_n}^{\tau+a_n} dH_n(x).$$

By Lemma 2 the right side tends to zero and we have

$$\lim_{n\to\infty} A_n = -\lim_{n\to\infty} \int_{-\tau}^{\tau} dH_n(x) = -\int_{-\tau}^{\tau} \frac{x^3}{1 + x^2}\, dM(x).$$

In a similar way we can prove that

$$\lim_{n\to\infty} C_n = \int_{|x|\geq\tau} \frac{x}{1 + x^2}\, dM(x)$$

and

$$\lim_{n\to\infty} E_n = -\int_{|x|>T} \frac{x}{1 + x^2}\, dM(x).$$

By use of the by now familiar device

$$\sum_{n=1}^{N} |r_n s_n| \leq \left(\max_{1\leq j\leq N} |r_j| \right) \sum_{n=1}^{N} |s_n|$$

we obtain

$$\lim_{n\to\infty} B_n = \lim_{n\to\infty} D_n = \lim_{n\to\infty} F_n = 0.$$

Since $|\lim_{n\to\infty} E_n|$ can be made arbitrarily small by taking T sufficiently large, we obtain (18), which concludes the proof of the theorem.

EXERCISES

1. There are many particles on a line, scattered at random, but on the average there is one per unit length. Each particle has a charge of $+1$ or -1, and there are roughly as many positively charged par-

ticles as there are negatively charged particles over any length of the line. The potential registered at a point P of a positively charged particle at Q is $1/|P - Q|$, and for a negatively charged particle it is $-1/|P - Q|$. The total potential at P is the sum of the potentials registered there by all the particles. Find the distribution function of the total potential at P.

2. This series of problems supplies some insight into Theorem 3.

(a) Let $A_1, A_2, \cdots, A_n$ be n independent events, and let $P(A_k) = p$ for $k = 1, 2, \cdots, n$, $0 < p < 1$. Let X denote the number of these events which occur. Find $P[X = k]$, $0 \leq k \leq n$.

(b) Let $Y_1, Y_2, \cdots, Y_n$ be n random variables. Let Z denote the number of these random variables which take values in $[a, b]$; that is, Z is the number of events among $\{[a \leq Y_j \leq b], 1 \leq j \leq n\}$ which occur. Show that Z is a random variable and in particular is a sum of indicators of events.

(c) Let $X_1, \cdots, X_n$ denote n independent random variables with the same distribution function F. Let N denote the number of these random variables which take values in $[a, b]$. Find the distribution $\{P[N = k], 0 \leq k \leq n\}$ of N in terms of F.

(d) For every n let $X_{n1}, \cdots, X_{nn}$ be n independent random variables with common distribution function F_n, and assume that

$$n(1 - F_n(x)) \to \lambda(x) \geq 0 \quad \text{as} \quad n \to \infty$$

for every $x > 0$. Let Z_n denote the number of the random variables among $X_{n1}, \cdots, X_{nn}$ which take values in $[a, b]$, where $0 < a < b$. Find the limiting distribution of Z_n.

(e) For every n let $X_{n1}, \cdots, X_{nn}$ be as in (d). Let

$$0 < a_1 < b_1 = a_2 < b_2 = a_3 < \cdots < b_{r-1} = a_r < b_r.$$

Let $Z_n^{(j)}$ denote the number of random variables among $X_{n1}, \cdots, X_{nn}$ which take values in $(a_j, b_j]$. Find the joint distribution of $Z_n^{(1)}, \cdots, Z_n^{(r)}$ in terms of F_n, and prove that, as $n \to \infty$,

$$\mathcal{L}(Z_n^{(1)}, \cdots, Z_n^{(r)}) \to \mathcal{L}(Z_1, \cdots, Z_r),$$

where for every i, $\mathcal{L}(Z_i)$ is $\mathcal{P}(\lambda(b_i) - \lambda(a_i))$ and $Z_1, \cdots, Z_r$ are independent.

(f) Suppose $U_1, \cdots, U_m$ are random variables, each of which takes values a and b, $a \neq b$, only, that is, $[U_j = a] \cup [U_j = b] = \Omega$. Let N_a denote the number of these random variables which equal a, and let N_b denote the number which equal b. Prove that $U_1 + \cdots + U_m = aN_a + bN_b$.

(g) Let $X_{n1}, \cdots, X_{nn}$ be as in Problem (d) with F_n purely discrete and having jumps only at 0, 1, 2, 3. Use Problems (e) and (f) to find the limiting distribution of $Z_n = X_{n1} + \cdots + X_{nn}$.

3. Attempt this series of problems only after successful completion of Exercise 2.

(a) For every n let $z_{n1}, \cdots, z_{nk_n}$ be k_n complex numbers such that: (α) $\max_{1 \le j \le k_n} |z_{nj}| \to 0$ as $n \to \infty$, (β) $\sum_{j=1}^{k_n} |z_{nj}|$ is bounded, and (γ) $k_n \to \infty$ as $n \to \infty$. Prove that $\lim_{n \to \infty} \prod_{j=1}^{k_n} (1 + z_{nj})$ exists if and only if $\lim_{n \to \infty} \sum_{j=1}^{k_n} z_{nj}$ exists, in which case

$$\lim_{n \to \infty} \prod_{j=1}^{k_n} (1 + z_{nj}) = \exp \lim_{n \to \infty} \sum_{j=1}^{k_n} z_{nj}.$$

(b) For every n let $A_{n1}, \cdots, A_{nk_n}$ be k_n independent events. Denote $p_{nj} = P(A_{nj})$. Let T_n denote the number of these events which occur. Find the characteristic function of T_n as a product of k_n terms.

(c) In (b) assume $\max_{1 \le j \le k_n} p_{nj} \to 0$ as $n \to \infty$. Prove that

$$\sum_{j=1}^{k_n} p_{nj} \to \lambda \ge 0 \qquad \text{as} \quad n \to \infty$$

if and only if the limiting distribution of T_n exists, it being $\mathcal{L}(0)$ if and only if $\lambda = 0$, and $\mathcal{P}(\lambda)$ if and only if $\lambda > 0$. (Actually $\mathcal{L}(0) = \mathcal{P}(0)$).

(d) Let $\mathfrak{X} = \{\{X_{nj}\}\}$ be an infinitesimal system, and let $0 < a_1 < b_1 = a_2 < b_2 = \cdots = a_r < b_r$, and let $Z_n^{(j)}$ denote the number of random variables among $X_{n1}, \cdots, X_{nk_n}$ which take values in $(a_j, b_j]$. Prove that

$$\mathcal{L}(Z_n^{(1)}, \cdots, Z_n^{(r)}) \to \mathcal{L}(Z^{(1)}, \cdots, Z^{(r)})$$

if and only if

$$\sum_{j=1}^{k_n} P[a_i < X_{nj} \le b_i] \to \lambda_i \ge 0 \qquad \text{as} \quad n \to \infty$$

for $1 \le i \le r$, in which case $Z^{(1)}, \cdots, Z^{(r)}$ are independent and $\mathcal{L}(Z^{(i)}) = \mathcal{P}(\lambda_i)$.

(e) Let $\mathfrak{X} = \{\{X_{nj}\}\}$ be an infinitesimal system of random variables such that for every n and j, X_{nj} takes values 0, 1, 2 only with probabilities $p_{nj}(0)$, $p_{nj}(1)$, and $p_{nj}(2)$, respectively. Prove, using previous exercises, that $\mathfrak{L}(\sum_{j=1}^{k_n} X_{nj})$ converges to a limit law if and only if $\sum_{j=1}^{k_n} p_{nj}(r) \to a_r \geq 0$ for $r = 1, 2$, that is,

$$\sum_{j=1}^{k_n} (-1 + F_{nj}(x)) \to M(x) \quad \text{for all} \quad x > 0.$$

Find the M-function and characteristic function of the limit law.

4. (This exercise provides an interpretation for Theorem 1.) Let $\mathfrak{X} = \{\{X_{nj}\}\}$ be an infinitesimal system, and let $\{c_n\}$ and $\{\{a_{nj}\}\}$ be as before. For every n let

$$\{Y_{nj}, X_{njr}, j = 1, 2, \cdots, k_n, r = 1, 2, \cdots\}$$

be independent random variables such that $\mathfrak{L}(Y_{nj}) = \mathcal{P}(1)$ for all n and all j and $\mathfrak{L}(X_{njr}) = \mathfrak{L}(X_{nj})$. Prove that the distribution function of $X_{n1} + \cdots + X_{nk_n} - c_n$ converges completely if and only if the distribution function of

$$\sum_{j=1}^{k_n} \left\{ \sum_{r=1}^{Y_{nj}} (X_{njr} - a_{nj}) + a_{nj} \right\} - c_n$$

converges completely, in which case both limit laws are the same.

6.6. Convergence to the Normal and Poisson Distributions.

In this section Theorem 3 of the previous section is used to obtain special limit theorems. In particular, conditions are given under which the distribution functions of a sequence of centered sums of independent random variables from an infinitesimal system converge to the normal distribution and to the Poisson distribution.

First, two lemmas are needed.

Lemma 1. If $0 \leq a_j \leq 1$ for $j = 1, 2, \cdots$, then

$$\prod_{j=1}^{n} (1 - a_j) \geq 1 - \sum_{j=1}^{n} a_j$$

for all n.

Proof: This is easily obtained by induction on n. It is true for $n = 1$. Assuming it true for any n, one obtains

$$\prod_{j=1}^{n+1} (1 - a_j) = \{\prod_{j=1}^{n} (1 - a_j)\} (1 - a_{n+1})$$

$$\geq \{1 - \sum_{j=1}^{n} a_j\} (1 - a_{n+1}) \geq 1 - \sum_{j=1}^{n+1} a_j,$$

which establishes the lemma.

Lemma 2. For every m, $m = 1, 2, \cdots, M$, let $\{ f_n^{(m)} \}$ be a sequence of real-valued functions defined over $(0, c)$, where $0 < c \leq \infty$. Suppose that for every m,

$$f_n^{(m)}(x) \to K^{(m)} \qquad \text{as} \quad n \to \infty$$

for every $x \in (0, c)$, where $K^{(m)}$ is a constant, finite or infinite. Then there exists a sequence $\{\epsilon_n\}$ in $(0, c)$ such that $0 < \epsilon_{n+1} < \epsilon_n$ for all n and $\epsilon_n \to 0$ as $n \to \infty$ which satisfies

$$f_n^{(m)}(\epsilon_n) \to K^{(m)} \qquad \text{as} \quad n \to \infty \quad \text{for} \quad m = 1, 2, \cdots, M.$$

Proof: We prove this only when $| K^{(m)} | < \infty$ for all m. The same proof holds if at least for one m, $K^{(m)} = \pm\infty$. Let $\epsilon > 0$. Then select $n_1 \geq 1$ such that

$$| f_n^{(m)}(\epsilon/2) - K^{(m)} | < \tfrac{1}{2} \qquad \text{for all} \quad n > n_1, 1 \leq m \leq M.$$

Let $n_2 > n_1$ be such that

$$| f_n^{(m)}(\epsilon/2^2) - K^{(m)} | < (\tfrac{1}{2})^2 \qquad \text{for all} \quad n > n_2, 1 \leq m \leq M.$$

In general, let $n_k > n_{k-1}$ be such that

$$| f_n^{(m)}(\epsilon/2^k) - K^{(m)} | < (\tfrac{1}{2})^k \qquad \text{for all} \quad n > n_k, 1 \leq m \leq M.$$

Let $n_0 = 0$; we define ϵ_n to be $\epsilon/2^k$ if $n_k < n \leq n_{k+1}$. Then for every m, if we denote $n_{k(r)}$ as the largest n_k which is *less than* r, we have

$$| f_r^{(m)}(\epsilon_r) - K^{(m)} | = | f_r^{(m)}(\epsilon/2^{k(r)}) - K^{(m)} |$$

$$< (\tfrac{1}{2})^{k(r)} \to 0 \qquad \text{as} \quad r \to \infty. \quad \text{Q.E.D.}$$

Remark 1. If F is an infinitely divisible distribution function determined by the triple (γ, σ^2, M), then F is normal with expectation μ and variance σ^2 if and only if $\gamma = \mu$ and $M \equiv 0$. This is immediate from Theorem 2 in Section 6.2.

Remark 2. If F is an infinitely divisible distribution function determined by the triple (γ, σ^2, M), then F has the Poisson distribution with expectation $\lambda > 0$ if and only if $\gamma = \lambda/2$, $\sigma^2 = 0$ and

$$
M(x) = \begin{cases} 0 & \text{if } x < 0 \text{ or } x > 1 \\ -\lambda & \text{if } 0 < x \leq 1. \end{cases}
$$

This observation immediately follows from Theorem 2 in Section 6.2.

Theorem 1 (Khinchine). Let $\mathfrak{X} = \{\{X_{nj}\}\}$ be an infinitesimal system of random variables, and let $Z_n = X_{n1} + \cdots + X_{nk_n}$. If $F_{Z_n} \xrightarrow{c} F$ as $n \to \infty$ where F is the distribution function of a nonconstant random variable, then F is a normal distribution function if and only if

$$
\sum_{j=1}^{k_n} \int_{|x|>\epsilon} dF_{nj}(x) \to 0 \qquad \text{as } n \to \infty \quad \text{for every } \epsilon > 0.
$$

Proof: We are given that $F_{Z_n} \xrightarrow{c} F$ as $n \to \infty$. By Theorem 3 in Section 6.5,

$$
\sum_{j=1}^{k_n} F_{nj}(x) \to M(x) \qquad \text{as } n \to \infty \quad \text{if } 0 > x \in Cont\ M,
$$

and

$$
\sum_{j=1}^{k_n} (F_{nj}(x) - 1) \to M(x) \qquad \text{as } n \to \infty \quad \text{if } 0 < x \in Cont\ M.
$$

By Remark 1 above, F is normal if and only if $M(-\epsilon) - M(\epsilon) = 0$ for every $\epsilon > 0$. Hence F is normal if and only if

$$
\sum_{j=1}^{k_n} \int_{|x|\geq\epsilon} dF_{nj}(x) = \sum_{j=1}^{k_n} (1 - F_{nj}(\epsilon) + F_{nj}(-\epsilon)) \to 0
$$

as $n \to \infty$. Q.E.D.

Theorem 2. Let $\mathfrak{X} = \{\{X_{nj}\}\}$ be an infinitesimal system of random variables, and let $Z_n = X_{n1} + \cdots + X_{nk_n}$. If $F_{Z_n} \overset{c}{\to} F$, then F is normal if and only if, for every $\epsilon > 0$,

$$(1) \qquad P[\max_{1 \leq j \leq k_n} |X_{nj}| \geq \epsilon] \to 0 \qquad \text{as} \quad n \to \infty.$$

Proof: We observe that

$$P[\max_{1 \leq j \leq k_n} |X_{nj}| \geq \epsilon] = 1 - P[\max_{1 \leq j \leq k_n} |X_{nj}| < \epsilon]$$

$$= 1 - \prod_{j=1}^{k_n} P[|X_{nj}| < \epsilon]$$

$$= 1 - \prod_{j=1}^{k_n} \left(1 - \int_{|x| \geq \epsilon} dF_{nj}(x)\right).$$

Hence condition (1) holds if and only if

$$(2) \qquad \prod_{j=1}^{k_n} \left(1 - \int_{|x| \geq \epsilon} dF_{nj}(x)\right) \to 1 \qquad \text{as} \quad n \to \infty.$$

By Lemma 1 and the fact that $1 - x \leq e^{-x}$ for $x \geq 0$, we obtain

$$1 - \sum_{j=1}^{k_n} \int_{|x| \geq \epsilon} dF_{nj}(x) \leq \prod_{j=1}^{k_n} \left(1 - \int_{|x| \geq \epsilon} dF_{nj}(x)\right)$$

$$\leq \exp\left\{-\sum_{j=1}^{k_n} \int_{|x| \geq \epsilon} dF_{nj}(x)\right\} \leq 1.$$

From this string of inequalities it follows that (2) holds if and only if

$$\sum_{j=1}^{k_n} \int_{|x| \geq \epsilon} dF_{nj}(x) \to 0 \qquad \text{as} \quad n \to \infty,$$

which by Theorem 1 is true if and only if F is normal. Q.E.D.

The most general theorem for convergence to a normal distribution is the following.

Theorem 3. Let $\mathfrak{X} = \{\{X_{nj}\}\}$ be a double sequence of row-wise independent random variables. In order that $\mathfrak{X}$ be an infinitesimal system and that there exist a sequence of constants $\{c_n\}$ such that $F_{Z_n} \overset{c}{\to} F$

as $n \to \infty$, where F is $\mathfrak{N}(0, 1)$ and $Z_n = X_{n1} + \cdots + X_{nk_n} - c_n$, it is necessary and sufficient that for all $\epsilon > 0$

(a) $\displaystyle \sum_{j=1}^{k_n} \int_{|x| \geq \epsilon} dF_{nj}(x) \to 0 \quad$ as $\quad n \to \infty$ and

(b) $\displaystyle \sum_{j=1}^{k_n} \left\{ \int_{|x| < \epsilon} x^2 \, dF_{nj}(x) - \left(\int_{|x| < \epsilon} x \, dF_{nj}(x) \right)^2 \right\} \to 1$

as $n \to \infty$.

Proof: We first prove that (a) and (b) are sufficient. From (a) we obtain

$$ 0 \leq \max_{1 \leq j \leq k_n} P[\, | X_{nj} | \geq \epsilon \,] \leq \sum_{j=1}^{k_n} \int_{|x| \geq \epsilon} dF_{nj}(x) \to 0 $$

as $n \to \infty$, which implies that $\mathfrak{X}$ is an infinitesimal system. Referring to Theorem 3 in Section 6.5 we note that (13) is satisfied by (a) when $M \equiv 0$ and that (14) with $\sigma^2 = 1$ is implied by (b). We take

$$ c_n = \sum_{j=1}^{k_n} a_{nj} $$

where

$$ a_{nj} = \int_{|x| < \tau} x \, dF_{nj}(x), \ \tau > 0, $$

and we take $\gamma = 0$. Thus by Theorem 3 of Section 6.5, $F_{Z_n} \xrightarrow{c} F$ where F is $\mathfrak{N}(0, 1)$. We prove next that conditions (a) and (b) are necessary. It is assumed that there exists a sequence $\{c_n\}$ such that $F_{Z_n} \xrightarrow{c} F$ where F is $\mathfrak{N}(0, 1)$, and $\mathfrak{X}$ is assumed to be an infinitesimal system. Thus in Theorem 3 in Section 6.5 we take $M \equiv 0$, $\gamma = 0$, $\sigma^2 = 1$ (because of Remark 1). Equation (13) in Section 6.5 clearly implies (a). We need only show that (13) and (14) imply (b). Let us denote

$$ J_n(\epsilon) = \sum_{j=1}^{k_n} \left(\int_{|x| < \epsilon} x^2 \, dF_{nj}(x) - \left(\int_{|x| < \epsilon} x \, dF_{nj}(x) \right)^2 \right), $$

and let $0 < \delta < \epsilon$. Then

$$ J_n(\epsilon) = J_n(\delta) + J_n(\delta, \epsilon) - 2L_n, $$

where

$$J_n(\delta, \epsilon) = \sum_{j=1}^{k_n} \left\{ \int_{\delta \le |x| < \epsilon} x^2 \, dF_{nj}(x) - \left(\int_{\delta \le |x| < \epsilon} x \, dF_{nj}(x) \right)^2 \right\},$$

and

$$L_n = \sum_{j=1}^{k_n} \left(\int_{|x| < \delta} x \, dF_{nj}(x) \right) \left(\int_{\delta \le |x| < \epsilon} x \, dF_{nj}(x) \right).$$

Now

$$0 \le J_n(\delta, \epsilon) \le \sum_{j=1}^{k_n} \int_{\delta \le |x| < \epsilon} x^2 \, dF_{nj}(x)$$

$$\le \epsilon^2 \sum_{j=1}^{k_n} \int_{\delta \le |x| < \epsilon} dF_{nj}(x) \to 0 \qquad \text{as} \quad n \to \infty$$

by (13) or (a). Also we have by (13) or (a) that

$$|L_n| \le \delta \epsilon \sum_{j=1}^{k_n} \int_{\delta \le |x| < \epsilon} dF_{nj}(x) \to 0 \qquad \text{as} \quad n \to \infty.$$

Hence

$$|J_n(\epsilon) - J_n(\delta)| \to 0 \qquad \text{as} \quad n \to \infty.$$

Thus

$$\limsup_{n \to \infty} J_n(\epsilon) = \limsup_{n \to \infty} J_n(\delta)$$

and

$$\liminf_{n \to \infty} J_n(\epsilon) = \liminf_{n \to \infty} J_n(\delta),$$

from which we deduce that $\limsup_{n \to \infty} J_n(\epsilon)$ and $\liminf_{n \to \infty} J_n(\epsilon)$ do not depend on the value of $\epsilon > 0$. Hence by (14) in Theorem 3 in Section 6.5, both $\limsup_{n \to \infty}$ and $\liminf_{n \to \infty}$ of the quantity

$$\sum_{j=1}^{k_n} \left\{ \int_{|x| < \epsilon} x^2 \, dF_{nj}(x) - \left(\int_{|x| < \epsilon} x \, dF_{nj}(x) \right)^2 \right\}$$

are equal to 1, thus implying (b). Q.E.D.

The following two theorems give conditions under which the partial sums of a sequence of independent random variables are asymptotically normal, that is, the distribution function of a linear function of the nth partial sum converges to a normal distribution.

Theorem 4 (S. N. Bernstein, W. Feller). Let $\{X_n\}$ be a sequence of independent random variables. In order that there exist two sequences of constants $\{A_n\}$, $\{B_n\}$, where $B_n > 0$ for all n, such that the distribution function of $(X_1 + \cdots + X_n)/B_n - A_n$ converges to that of the $\mathfrak{N}(0, 1)$ distribution and that $\{\{X_{nj}\}\}$ be an infinitesimal system, where $X_{nj} = X_j/B_n$, $1 \le j \le n$, it is necessary and sufficient that there exist a sequence of real numbers $\{C_n\}$ such that $C_n \to \infty$ as $n \to \infty$, and such that

$$(\alpha) \qquad \sum_{j=1}^{n} \int_{|x|>C_n} dF_j(x) \to 0 \qquad \text{as} \quad n \to \infty, \text{ and}$$

$$(\beta) \qquad \frac{1}{C_n{}^2} \sum_{j=1}^{n} \left\{ \int_{|x|<C_n} x^2 \, dF_j(x) - \left(\int_{|x|<C_n} x \, dF_j(x) \right)^2 \right\} \to \infty$$

as $n \to \infty$, where F_k is the distribution function of X_k.

Proof: We first prove that the condition is necessary. Accordingly we assume that $F_{Z_n} \xrightarrow{c} F$, where F is $\mathfrak{N}(0, 1)$, $\{\{X_{nj}\}\}$ as defined above is an infinitesimal system, and where $Z_n = (X_1 + \cdots + X_n)/B_n - A_n$. At least one random variable, say, X_k, is not zero with probability one. Hence, since $\{\{X_{nj}\}\}$ is an infinitesimal system it follows that

$$X_k/B_n \xrightarrow{P} 0 \qquad \text{as} \quad n \to \infty,$$

and hence $B_n \to +\infty$. Let F_{nj} be the distribution function of X_{nj}. Thus $F_{nj}(x) = F_j(B_n x)$. By Theorem 3 we obtain

$$(\alpha') \qquad \sum_{j=1}^{n} \int_{|x|>\epsilon B_n} dF_j(x) \to 0 \qquad \text{as} \quad n \to \infty, \text{ and}$$

$$(\beta') \qquad \frac{1}{B_n{}^2} \sum_{j=1}^{n} \left\{ \int_{|x|<\epsilon B_n} x^2 \, dF_j(x) - \left(\int_{|x|<\epsilon B_n} x \, dF_j(x) \right)^2 \right\} \to 1$$

as $n \to \infty$ for every $\epsilon > 0$. By Lemma 2 we can select a sequence $\{\epsilon_n\}$ of positive numbers such that $\epsilon_n \to 0$ as $n \to \infty$ and such that

$$(3) \qquad \epsilon_n B_n \to \infty \qquad \text{as} \quad n \to \infty,$$

$$(4) \qquad \sum_{j=1}^{n} \int_{|x|>\epsilon_n B_n} dF_j(x) \to 0 \qquad \text{as} \quad n \to \infty,$$

and

(5) $\displaystyle \frac{1}{B_n{}^2} \sum_{j=1}^{n} \left\{ \int_{|x|<\epsilon_n B_n} x^2 \, dF_j(x) - \left(\int_{|x|<\epsilon_n B_n} x \, dF_j(x) \right)^2 \right\} \to 1.$

Let $C_n = \epsilon_n B_n$. Then (3) implies that $C_n \to \infty$, and (4) and (5) yield conditions (α) and (β), respectively, thus proving these conditions necessary. We now prove that conditions (α) and (β) are sufficient. Let us denote $B_n = +\sqrt{B_n{}^2}$, where

$$B_n{}^2 = \sum_{j=1}^{n} \left\{ \int_{|x|\leq C_n} x^2 \, dF_j(x) - \left(\int_{|x|\leq C_n} x \, dF_j(x) \right)^2 \right\}.$$

Condition (β) implies that $B_n{}^2/C_n{}^2 \to \infty$ or $C_n/B_n \to 0$ as $n \to \infty$. Let $\epsilon > 0$ be arbitrary. Then for sufficiently large n

$$\sum_{j=1}^{n} \int_{|x|>C_n} dF_j(x) \geq \sum_{j=1}^{n} \int_{|x|>\epsilon B_n} dF_j(x).$$

Condition (α) thus implies that

(6) $\displaystyle \sum_{j=1}^{n} \int_{|x|>\epsilon B_n} dF_j(x) \to 0 \qquad \text{as} \quad n \to \infty.$

It is seen that (6) implies that $\{\{X_j/B_n, 1 \leq j \leq n\}\}$ is an infinitesimal system which satisfies condition (a) in Theorem 3. Next we observe that for sufficiently large n that $0 < C_n < \epsilon B_n$, and thus if we denote

$$L_n = \frac{1}{B_n{}^2} \sum_{j=1}^{n} \left(\int_{|x|<\epsilon B_n} x^2 \, dF_j(x) - \left(\int_{|x|<\epsilon B_n} x \, dF_j(x) \right)^2 \right),$$

we may write $L_n = K_n + M_n + N_n$, where

$$K_n = \frac{1}{B_n{}^2} \sum_{j=1}^{n} \left(\int_{|x|\leq C_n} x^2 \, dF_j(x) - \left(\int_{|x|\leq C_n} x \, dF_j(x) \right)^2 \right),$$

$$M_n = \frac{1}{B_n{}^2} \sum_{j=1}^{n} \left(\int_{C_n<|x|<\epsilon B_n} x^2 \, dF_j(x) \right.$$

$$\left. - \left(\int_{C_n<|x|<\epsilon B_n} x \, dF_j(x) \right)^2 \right),$$

and

$$N_n = -\frac{2}{B_n^2} \sum_{j=1}^{n} \left(\int_{|x| \le C_n} x\, dF_j(x) \right) \left(\int_{C_n < |x| < \epsilon B_n} x\, dF_j(x) \right).$$

But

$$0 \le M_n \le \frac{1}{B_n^2} \sum_{j=1}^{n} \int_{C_n < |x| < \epsilon B_n} x^2\, dF_j(x)$$

$$\le \epsilon^2 \sum_{j=1}^{n} \int_{|x| > C_n} dF_j(x),$$

and thus by condition (α), $M_n \to 0$ as $n \to \infty$. Also

$$|N_n| \le 2\epsilon(C_n/B_n) \sum_{j=1}^{n} \int_{|x| > C_n} dF_j(x),$$

and by (α) and the fact that $C_n/B_n \to 0$ we obtain that $N_n \to 0$. Since $K_n \equiv 1$, we obtain $L_n \to 1$ as $n \to \infty$, and thus by Theorem 3 we have shown that conditions (α) and (β) are sufficient for normal convergence. Q.E.D.

The following theorem is the most general theorem for normal convergence for normed centered partial sums of a sequence of independent random variables with finite second moments.

Theorem 5 (Lindeberg, Feller). Let $\{X_n\}$ be a sequence of independent random variables with finite variances, let F_j denote the distribution function of X_j, let $S_n = X_1 + \cdots + X_n$, let

$$X_{nj} = (X_j - EX_j)/\sqrt{\mathrm{Var}\ S_n},$$

and let $Z_n = \sum_{j=1}^{n}(X_j - EX_j)/\sqrt{\mathrm{Var}\ S_n}$. In order that $F_{Z_n} \xrightarrow{c} F$, where F is $\mathfrak{N}(0, 1)$, and that $\{\{X_{nj},\ 1 \le j \le n\}\}$ be an infinitesimal system it is necessary and sufficient that

$$\frac{1}{\mathrm{Var}\ S_n} \sum_{j=1}^{n} \int_{|x| \ge \epsilon\sqrt{\mathrm{Var}\ S_n}} x^2\, dF_j(x + EX_j) \to 0$$

as $n \to \infty$ for every $\epsilon > 0$.

Proof: Let us denote, for $\epsilon > 0$,

$$O_n = \frac{1}{\text{Var } S_n} \sum_{j=1}^{n} \int_{|x| \geq \epsilon \sqrt{\text{Var } S_n}} x^2 \, dF_j(x + EX_j),$$

$$I_n = \frac{1}{\text{Var } S_n} \sum_{j=1}^{n} \int_{|x| < \epsilon \sqrt{\text{Var } S_n}} x^2 \, dF_j(x + EX_j).$$

Clearly $O_n + I_n = 1$ for all n. We first prove that the condition in the theorem is sufficient. In the expression for O_n, we replace x by $x - EX_j$ and obtain

$$O_n \geq \epsilon^2 \sum_{j=1}^{n} \int_{|x - EX_j| \geq \epsilon \sqrt{\text{Var } S_n}} dF_j(x) \geq 0.$$

Since $O_n \to 0$ as $n \to \infty$, then condition (a) of Theorem 3 is satisfied. Since $O_n \to 0$ and $O_n + I_n = 1$, then $I_n \to 1$ as $n \to \infty$, or

$$I_n = \sum_{j=1}^{n} \int_{|x| < \epsilon} x^2 \, dF_{nj}(x) \to 1 \qquad \text{as} \quad n \to \infty,$$

where F_{nj} is the distribution function of

$$X_{nj} = (X_j - EX_j)/\sqrt{\text{Var } S_n}.$$

In order to show that condition (b) of Theorem 3 is satisfied we need only prove that if we denote

$$Q_n = \sum_{j=1}^{n} \left(\int_{|x| < \epsilon} x \, dF_{nj}(x) \right)^2,$$

then $Q_n \to 0$ as $n \to \infty$ for every $\epsilon > 0$. Since $\int x \, dF_{nj}(x) = 0$, then

$$\left| \int_{|x| < \epsilon} x \, dF_{nj}(x) \right| = \left| \int_{|x| \geq \epsilon} x \, dF_{nj}(x) \right|.$$

We note that

$$0 \leq \sum_{j=1}^{n} \left| \int_{|x| \geq \epsilon} x \, dF_{nj}(x) \right| \leq \sum_{j=1}^{n} \int_{|x| \geq \epsilon} |x| \, dF_{nj}(x)$$

$$\leq \frac{1}{\epsilon} \sum_{j=1}^{n} \int_{|x| \geq \epsilon} x^2 \, dF_{nj}(x) = \frac{O_n}{\epsilon} \to 0 \qquad \text{as} \quad n \to \infty.$$

From this we obtain

$$\max_{1 \leq j \leq n} \left| \int_{|x| \geq \epsilon} x \, dF_{nj}(x) \right| \leq \sum_{j=1}^{n} \left| \int_{|x| \geq \epsilon} x \, dF_{nj}(x) \right| \to 0$$

as $n \to \infty$, and hence

$$\max_{1 \leq j \leq n} \left| \int_{|x| < \epsilon} x \, dF_{nj}(x) \right| \to 0 \qquad \text{as} \quad n \to \infty.$$

In addition,

$$\sum_{j=1}^{n} \left| \int_{|x| < \epsilon} x \, dF_{nj}(x) \right| = \sum_{j=1}^{n} \left| \int_{|x| \geq \epsilon} x \, dF_{nj}(x) \right| \to 0$$

as $n \to \infty$. From these last two relations we obtain

$$Q_n \leq \left\{ \max_{1 \leq j \leq n} \left| \int_{|x| < \epsilon} x \, dF_{nj}(x) \right| \right\} \sum_{j=1}^{n} \left| \int_{|x| < \epsilon} x \, dF_{nj}(x) \right| \to 0$$

as $n \to \infty$. Thus the condition is sufficient. We now prove that the condition is necessary. Assuming $F_{Z_n} \xrightarrow{c} F$, where F is $\mathfrak{N}(0, 1)$, we have by Theorem 3 that, for every $\epsilon > 0$,

(a) $$\sum_{j=1}^{n} \int_{|x| \geq \epsilon} dF_{nj}(x) \to 0 \qquad \text{as} \quad n \to \infty, \text{ and}$$

(b) $$\sum_{j=1}^{n} \left\{ \int_{|x| < \epsilon} x^2 \, dF_{nj}(x) - \left(\int_{|x| < \epsilon} x \, dF_{nj}(x) \right)^2 \right\} \to 1$$

as $n \to \infty$. Condition (b) implies that for every $\zeta > 0$,

$$\sum_{j=1}^{n} \int_{|x| < \epsilon} x^2 \, dF_{nj}(x) > 1 - \zeta$$

for sufficiently large n. But

$$\sum_{j=1}^{n} \int_{-\infty}^{\infty} x^2 \, dF_{nj}(x) = 1$$

for all n. Hence

$$\sum_{j=1}^{n} \int_{|x| \geq \epsilon} x^2 \, dF_{nj}(x) < \zeta$$

for sufficiently large n. Q.E.D.

The following corollary is what is generally referred to by statisticians as "the central limit theorem."

Corollary to Theorem 5. Let $\{X_n\}$ be a sequence of independent identically distributed random variables with common expectation μ and variance $\sigma^2 > 0$. Then $F_{Z_n} \xrightarrow{c} F$, where F is $\mathfrak{N}(0, 1)$ and

$$Z_n = (X_1 + \cdots + X_n - n\mu)/\sqrt{n\sigma^2}.$$

Proof: Let us denote $X_{nj} = (X_j - \mu)/\sqrt{n\sigma^2}$, F_{nj} the distribution function of X_{nj}, and F_1 the distribution function of X_1. Then $EX_{nj} = 0$, $\operatorname{Var} X_{nj} = 1/n$, and

$$0 \le \sum_{j=1}^{n} \int_{|x| \ge \epsilon} x^2 \, dF_{nj}(x) = n \int_{|x| \ge \epsilon} x^2 \, dF_{nj}(x)$$

$$= n \int_{|x-\mu| \ge \epsilon\sqrt{n\sigma^2}} \frac{(x - \mu)^2}{n\sigma^2} \, dF_1(x) \to 0 \qquad \text{as} \quad n \to \infty$$

since $EX_1^2 < \infty$. Thus by Theorem 5 we obtain the conclusion.

The following theorem gives necessary and sufficient conditions for convergence to the Poisson distribution.

Theorem 6. Let $\mathfrak{X} = \{\{X_{nj}\}\}$ be an infinitesimal system of random variables, and let $Z_n = X_{n1} + \cdots + X_{nk_n}$. In order that $F_{Z_n} \xrightarrow{c} F$, where F is the Poisson distribution function with expectation $\lambda > 0$, it is necessary and sufficient that, for $0 < \epsilon < \frac{1}{2}$,

(a) $\quad \sum_{j=1}^{k_n} \int_{-\infty}^{-\epsilon} + \int_{\epsilon}^{1-\epsilon} + \int_{1+\epsilon}^{\infty} dF_{nj}(x) \to 0$,

(b) $\quad \sum_{j=1}^{k_n} \int_{|x-1|<\epsilon} dF_{nj}(x) \to \lambda$,

(c) $\quad \sum_{j=1}^{k_n} \int_{|x|<\epsilon} x \, dF_{nj}(x) \to 0$, and

(d) $\quad \sum_{j=1}^{k_n} \left\{ \int_{|x|<\epsilon} x^2 \, dF_{nj}(x) - \left(\int_{|x|<\epsilon} x \, dF_{nj}(x) \right)^2 \right\} \to 0$

as $n \to \infty$.

Proof: We first recall that the characteristic function of the Poisson distribution with expectation $\lambda > 0$ is

$$f(u) = \exp \lambda (e^{iu} - 1)$$

$$= \exp \left\{ \frac{iu\lambda}{2} + \int_{-\infty}^{\infty} \left(e^{iux} - 1 - \frac{iux}{1 + x^2} \right) dM(x) \right\},$$

where

$$M(x) = \begin{cases} 0 & \text{if} \quad x < 0 \quad \text{or} \quad x \geq 1 \\[2mm] -\lambda & \text{if} \quad 0 < x < 1 \end{cases}$$

Let us denote

$$M_n(x) = \begin{cases} \displaystyle\sum_{j=1}^{k_n} F_{nj}(x) & \text{if} \quad x < 0 \\[6mm] \displaystyle\sum_{j=1}^{k_n} (F_{nj}(x) - 1) & \text{if} \quad x > 0, \end{cases}$$

and let F be the distribution function determined by f. By Theorem 3 of Section 6.5, $F_{Z_n} \xrightarrow{c} F$ if and only if

(7) $$M_n(x) \to M(x) \qquad \text{as} \quad n \to \infty \quad \text{if} \quad x \neq 0, 1,$$

(8) $$\lim_{\epsilon \downarrow 0} \begin{Bmatrix} \limsup \\ \liminf \end{Bmatrix}_{n \to \infty} \sum_{j=1}^{k_n} \left\{ \int_{|x| < \epsilon} x^2 \, dF_{nj}(x) - \left(\int_{|x| < \epsilon} x \, dF_{nj}(x) \right)^2 \right\} = 0,$$

and

(9) $$\sum_{j=1}^{k_n} \int_{|x| < \tau} x \, dF_{nj}(x) \to \frac{\lambda}{2} + \int_{|x| < \tau} \frac{x^3}{1 + x^2} \, dM(x)$$

$$- \int_{|x| > \tau} \frac{x}{1 + x^2} \, dM(x)$$

for $\tau > 0$, $\tau \neq 1$. Conditions (a) and (b) are easily seen to be equivalent to (7). Condition (d) implies (8) by brief inspection. Since we restrict $0 < \epsilon < \frac{1}{2}$, then in the same manner as was done in the proof of Theorem

3, we obtain by (7) and (8) that

$$
\lim \begin{Bmatrix} \lim\sup \\ \lim\inf \end{Bmatrix}_{n\to\infty} \sum_{j=1}^{k_n} \left\{ \int_{|x|<\epsilon} x^2 \, dF_{nj}(x) - \left(\int_{|x|<\epsilon} x \, dF_{nj}(x) \right)^2 \right\}
$$

does not depend on ϵ and therefore implies condition (d). We need only show that (c) and (9) are equivalent under conditions (7) and (8). If $0 < \tau < 1$, then by (7) and (9) we have

$$
\sum_{j=1}^{k_n} \int_{|x|<\tau} x \, dF_{nj}(x) \to \tfrac{1}{2}\lambda + 0 - \tfrac{1}{2}\lambda = 0 \qquad \text{as} \quad n \to \infty,
$$

which yields (c). Conversely (7) and (c) are easily seen to imply (9) when $0 < \tau < 1$, and thus when $0 < \tau < 1$, (c) and (9) are equivalent under (7) and (8). In case $\tau > 1$, we write

$$
\sum_{j=1}^{k_n} \int_{|x|<\tau} x \, dF_{nj}(x) = A_n + B_n,
$$

where

$$
A_n = \sum_{j=1}^{k_n} \int_{|x|<\epsilon} x \, dF_{nj}(x)
$$

and

$$
B_n = \sum_{j=1}^{k_n} \int_{\epsilon \le |x| < \tau} x \, dF_{nj}(x).
$$

Also, for $\tau > 1$, one easily computes that

$$
\tfrac{1}{2}\lambda + \int_{|x|<\tau} \frac{x^3}{1+x^2} \, dM(x) - \int_{|x|>\tau} \frac{x}{1+x^2} \, dM(x) = \lambda.
$$

If (a)–(d) are true, then $A_n \to 0$ and $B_n \to \lambda$, thus implying (9). If (7), (8) and (9) are true, then by (7), $B_n \to \lambda$, and thus the above computation and (9) imply $A_n \to 0$ which gives (c). Thus we have proved (c) is equivalent to (9) under (7) and (8) when $\tau > 1$. Q.E.D.

Exercises

1. Let $\mathfrak{X} = \{\{X_{n,1}, \cdots, X_{n,k_n}\}\}$ be a double sequence of random variables which are row-wise independent. Prove that if

$$P[\max_{1 \leq j \leq k_n} | X_{nj} | \geq \epsilon] \to 0 \qquad \text{as} \quad n \to \infty$$

for every $\epsilon > 0$, then $\mathfrak{X}$ is an infinitesimal system.

2. For every $t \in [0, \infty)$, let X_t be a random variable. The set of random variables $\{X_t, 0 \leq t < \infty\}$ is assumed to have the following property: for every n and every $0 \leq t_0 < t_1 < \cdots < t_n$, the random variables $\{X_{t_j} - X_{t_{j-1}}, 1 \leq j \leq n\}$ are independent. Let $t > 0$ and let

$$0 = t_{n,0} < t_{n,1} < \cdots < t_{n,n} = t$$

be such that $\max_{1 \leq j \leq n}(t_{nj} - t_{n,j-1}) \to 0$ as $n \to \infty$. Prove that if

$$P[\max_{1 \leq j \leq n} | X_{t_{n,j}} - X_{t_{n,j-1}} | \geq \epsilon] \to 0 \qquad \text{as} \quad n \to \infty$$

for every $\epsilon > 0$, then the distribution of $X(t) - X(0)$ is normal.

3. Let $\{X_n\}$ be a sequence of independent, identically distributed random variables with common finite expectation μ and variance $0 < \sigma^2 < \infty$. Denote $\bar{X}_n = (X_1 + \cdots + X_n)/n$, $s_n^2 = (1/(n-1)) \sum_{j=1}^n (X_j - \bar{X}_n)^2$, $s_n = +\sqrt{s_n^2}$, and $Z_n = (\bar{X}_n - \mu)/(s_n/\sqrt{n})$. Prove that $F_{Z_n} \xrightarrow{c} F$ where F is $\mathfrak{N}(0, 1)$.

CHAPTER 7

Conditional Expectation
and Martingale Theory

7.1. Conditional Expectation

Conditional expectation is one of the fundamental notions in probability theory and is a most frequently used tool. In this section the definition of conditional expectation is given, and some of its fundamental properties are established. In a subsequent section of this chapter this basic tool is used to obtain martingale convergence theorems.

Conditional expectation could not be properly defined until the Radon-Nikodym theorem was stated and proved in its abstract measure-theoretic setting. Here we recall the statement of that theorem upon which we shall depend. Let φ be a finite signed measure (that is, a difference of two finite measures) over a finite measure space $(\mathfrak{X}, \mathfrak{M}, \mu)$, and assume that φ is absolutely continuous with respect to μ, that is, $M \in \mathfrak{M}$ and $\mu(M) = 0$ imply $\varphi(M) = 0$. Then there exists an $\mathfrak{M}$-measurable function f, uniquely determined except over an $\mathfrak{M}$-measurable set of μ-measure zero, such that $\varphi(M) = \int_M f \, d\mu$ for all $M \in \mathfrak{M}$.

The definition of conditional expectation can be developed as follows. Let $(\Omega, \mathfrak{a}, P)$ be a probability space, and let $\mathfrak{B}$ be a sub-sigma field of events: $\mathfrak{B} \subset \mathfrak{a}$. Let X be a random variable with finite expectation. We define a set function φ over $\mathcal{B}$ by

$$\varphi(B) = \int_B X \, dP = E(XI_B) \qquad \text{for all} \quad B \in \mathcal{B}.$$

Clearly φ is a signed measure over $(\Omega, \mathcal{B})$ and is absolutely continuous with respect to P. By the Radon-Nikodym theorem stated above there exists a $\mathcal{B}$-measurable random variable which we denote by $E(X \mid \beta)$, uniquely determined except over a $\mathcal{B}$-measurable event of probability zero, such that

$$\varphi(B) = \int_B E(X \mid \mathcal{B}) \, dP \qquad \text{for all} \quad B \in \mathcal{B}.$$

The random variable $E(X \mid \mathcal{B})$ is called the conditional expectation of X given the sigma field $\mathcal{B}$. A formal definition of conditional expectation can now be given.

Definition: Let X be a random variable with finite expectation, and let $\mathcal{B}$ be a sub-sigma field of $\mathcal{A}$. The conditional expectation of X given $\mathcal{B}$, $E(X \mid \mathcal{B})$, is a $\mathcal{B}$-measurable random variable, uniquely determined except over an event of probability zero, which satisfies

$$\int_B X \, dP = \int_B E(X \mid \mathcal{B}) \, dP \qquad \text{for all} \quad B \in \mathcal{B}.$$

Remark 1. If $X \in L_1(\Omega, \mathcal{A}, P)$, and if $\mathcal{B}$ is a sub-sigma field of $\mathcal{A}$, then $EX = E(E(X \mid \mathcal{B}))$. This follows immediately from the above identity in the case $B = \Omega$.

Remark 2. If $X \in L_1(\Omega, \mathcal{A}, P)$, and if $\mathcal{B} = \{\phi, \Omega\}$, then

$$E(X \mid \mathcal{B})(\omega) = EX$$

for *all* $\omega \in \Omega$. Indeed, the constant EX satisfies the identity in the above definition. The conclusion of this remark holds for all $\omega \in \Omega$ because of the fact that the only event in $\mathcal{B}$ of probability zero is ϕ.

Remark 3. If $X \in L_1(\Omega, \mathcal{A}, P)$, and if $\mathcal{B}$ is a sub-sigma field of $\mathcal{A}$ consisting only of events of probability zero or one, then $E(X \mid \mathcal{B}) = EX$ except over a $\mathcal{B}$-measurable event of probability zero. This remark is obvious.

Remark 4. If X and Y are in $L_1(\Omega, \, \mathcal{C}, \, P)$, if a, b are constants, and if $\mathcal{B}$ is a sub-sigma field of $\mathcal{C}$, then

$$E(aX + bY \mid \mathcal{B}) = aE(X \mid \mathcal{B}) + bE(Y \mid \mathcal{B}) \text{ a.s.,}$$

and if $X \leq Y$ a.s., then $E(X \mid \mathcal{B}) \leq E(Y \mid \mathcal{B})$ a.s.

Remark 5. If $X \in L_1(\Omega, \, \mathcal{C}, \, P)$, if $\mathcal{B}$ is a sub-sigma field of $\mathcal{C}$, and if $E(X \mid \mathcal{B}) \leq EX$ a.s., then $E(X \mid \mathcal{B}) = EX$ a.s.

In case the conditioning sigma field is the sigma field generated by a class of random variables, we shall replace it in the notation for conditional expectation by those random variables.

Definition: If $X \in L_1(\Omega, \, \mathcal{C}, \, P)$, and if $\{U, \, V, \, \cdots\}$ is a collection of random variables, then we define

$$E(X \mid U, \, V, \, \cdots) = E(X \cdot \mid \sigma\{U, \, V, \, \cdots\}).$$

The following theorems provide basic properties of conditional expectation.

Theorem 1. If $X \in L_1(\Omega, \, \mathcal{C}, \, P)$, and if Y is a random variable, then $E(X \mid Y)$ is a Borel-measurable function of Y.

Proof: This is an immediate consequence of Theorem 1 in Section 1.3 and the definition of $E(X \mid Y)$.

Theorem 2. If $\mathcal{B}$ is the sub-sigma field of $\mathcal{C}$ generated by the disjoint events $\{B_0, B_1, B_2, \cdots\}$, where $P(B_0) = 0$, $P(B_n) > 0$ for $n = 1, 2, \cdots$, and such that $\Omega = \cup_{n=0}^{\infty} B_n$, then for every $\omega \in B_n$, $E(X \mid \mathcal{B})(\omega) = E(XI_{B_n})/P(B_n)$ for $n = 1, 2, \cdots$.

Proof: Let $W = \sum_{j=1}^{\infty} \{E(XI_{B_j})/P(B_j)\} I_{B_j}$. If $B \in \mathcal{B}$, then there is a finite or infinite sequence $\{n_j\}$ of positive integers such that

$$B = \underset{j}{\cup} B_{n_j}.$$

Then

$$\int_B W\,dP = \sum_j \int_{B_{n_j}} W\,dP = \sum_j \int I_{B_{n_j}} W\,dP$$

$$= \sum_j E(XI_{B_{n_j}}) = E(XI_B) = \int_B X\,dP.$$

Since W is $\mathcal{B}$-measurable, then $W = E(X \mid \mathcal{B})$ except over a $\mathcal{B}$-measurable event of probability zero. But B_0 is the only event in $\mathcal{B}$ of probability zero! Q.E.D.

The following theorem is one of the two most frequently used properties of conditional expectation.

Theorem 3. If $X \in L_1(\Omega, \mathcal{A}, P)$, if $\mathcal{B}$ and $\mathcal{C}$ are sub-sigma fields of $\mathcal{A}$, and if $\mathcal{B} \subset \mathcal{C}$, then $E(X \mid \mathcal{B}) = E(E(X \mid \mathcal{C}) \mid \mathcal{B})$ a.s.

Proof: Since both sides are $\mathcal{B}$-measurable, one need only verify that

$$\int_B E(X \mid \mathcal{B})\,dP = \int_B E(E(X \mid \mathcal{C}) \mid \mathcal{B})\,dP \qquad \text{for all} \quad B \in \mathcal{B}.$$

By the definition, for $B \in \mathcal{B}$,

$$\int_B E(X \mid \mathcal{B})\,dP = \int_B X\,dP,$$

and since $\mathcal{B} \subset \mathcal{C}$, then $B \in \mathcal{C}$, so

$$\int_B E(E(X \mid \mathcal{C}) \mid \mathcal{B})\,dP = \int_B E(X \mid \mathcal{C})\,dP = \int_B X\,dP,$$

from which the conclusion follows.

The following theorem is the second of the two most frequently used properties of conditional expectation.

Theorem 4. If X and Y are random variables, if Y and XY have finite expectations, if $\mathcal{B}$ is a sub-sigma field of $\mathcal{A}$, and if X is $\mathcal{B}$-measurable, then $E(XY \mid \mathcal{B}) = XE(Y \mid \mathcal{B})$ a.s.

Proof: Since both sides are $\mathcal{B}$-measurable we need only prove that

$$\int_B E(XY \mid \mathcal{B})\, dP = \int_B XE(Y \mid \mathcal{B})\, dP \qquad \text{for all} \quad B \in \mathcal{B}.$$

By the definition of conditional expectation,

$$\int_B E(XY \mid \mathcal{B})\, dP = \int_B XY\, dP \qquad \text{for all} \quad B \in \mathcal{B}.$$

Let us first consider the case where X is discrete, in particular,

$$X = \sum_{n=1}^{N} b_n I_{B_n},$$

where N is finite and $B_n \in \mathcal{B}$ for $1 \leq n \leq N$. In this case, for $B \in \mathcal{B}$, then $BB_n \in \mathcal{B}$, and

$$\int_B XE(Y \mid \mathcal{B})\, dP = \int_B \left(\sum_{n=1}^{N} b_n I_{B_n}\right) E(Y \mid \mathcal{B})\, dP$$

$$= \sum_{n=1}^{N} b_n \int_{BB_n} E(Y \mid \mathcal{B})\, dP$$

$$= \sum_{n=1}^{N} b_n \int_{BB_n} Y\, dP = \int_B \left(\sum_{n=1}^{N} b_n I_{B_n}\right) Y\, dP$$

$$= \int_B XY\, dP.$$

Hence the theorem is true for X a discrete random variable which takes a finite number of values with positive probabilities. Let us denote $U^+ = UI_{[U \geq 0]}$ and $U^- = -UI_{[U < 0]}$ for U a random variable. It follows from Remark 4 above that

$$XE(Y \mid \mathcal{B}) = X^+ E(Y^+ \mid \mathcal{B}) - X^+ E(Y^- \mid \mathcal{B})$$
$$- X^- E(Y^+ \mid \mathcal{B}) + X^- E(Y^- \mid \mathcal{B}) \text{ a.s.}$$

The proof will be accomplished if we can prove, say, that

$$E(X^+ Y^+ \mid \mathcal{B}) = X^+ E(Y^+ \mid \mathcal{B}) \text{ a.s.}$$

Accordingly, let $\{X_n^+\}$ be a monotone nondecreasing sequence of $\mathcal{B}$-measurable discrete nonnegative random variables, each of which has a finite range, such that $X_n{}^+ \to X^+$ a.s. Then easily $\{X_n{}^+Y^+\}$ is nondecreasing and nonnegative and converges almost surely to X^+Y^+. (Note that $X^+Y^+ \in L_1(\Omega, \mathcal{Q}, P)$ since $0 \le X^+Y^+ \le |XY|$.) Now by the monotone convergence theorem and the special case proved above we have for every $B \in \mathcal{B}$

$$\int_B X^+E(Y^+ \mid \mathcal{B}) \, dP = \lim_{n \to \infty} \int_B X_n{}^+E(Y^+ \mid \mathcal{B}) \, dP$$

$$= \lim_{n \to \infty} \int_B E(X_n{}^+Y^+ \mid \mathcal{B}) \, dP$$

$$= \lim_{n \to \infty} \int_B X_n{}^+Y^+ \, dP = \int_B X^+Y^+ \, dP$$

$$= \int_B E(X^+Y^+ \mid \mathcal{B}) \, dP,$$

and thus $E(X^+Y^+ \mid \mathcal{B}) = X^+E(Y^+ \mid \mathcal{B})$ a.s. Q.E.D.

Theorem 5. If X and Y are independent random variables, and if $X \in L_1(\Omega, \mathcal{Q}, P)$, then $E(X \mid Y) = EX$ a.s. (Equivalently, if

$$X \in L_1(\Omega, \mathcal{Q}, P),$$

if $\mathcal{B}$ is a sub-sigma field of $\mathcal{Q}$, and if $\sigma\{X\}$ and $\mathcal{B}$ are independent classes of events, then $E(X \mid \mathcal{B}) = EX$ a.s.).

Proof: If $B \in \sigma\{Y\}$, then by Theorem 1 of Section 1.3, I_B is a Borel-measurable function of Y, and by Theorem 3 in Section 3.2, X and I_B are independent. Hence for every $B \in \sigma\{Y\}$,

$$\int_B E(X \mid Y) \, dP = \int_B X \, dP = \int_B I_B X \, dP = (EX)(E(I_B))$$

$$= \int_B (EX) \, dP.$$

The constant EX is trivially measurable with respect to $\sigma\{Y\}$, and thus by uniqueness $E(X \mid Y) = EX$ a.s. Q.E.D.

Theorem 6. If $\mathcal{B}$ and $\mathcal{C}$ are two sub-sigma fields of $\mathcal{C}$, if $\mathcal{B} \subset \mathcal{C}$ and if $X \in L_1(\Omega, \mathcal{C}, P)$, then $E(X \mid \mathcal{B}) = E(E(X \mid \mathcal{B}) \mid \mathcal{C})$ a.s.

Proof: Since $E(X \mid \mathcal{B})$ is $\mathcal{B}$-measurable, it is $\mathcal{C}$-measurable, so by Theorem 4

$$E(E(X \mid \mathcal{B}) \mid \mathcal{C}) = E(X \mid \mathcal{B})E(1 \mid \mathcal{C}) = E(X \mid \mathcal{B}) \text{ a.s.}$$

We shall subsequently need a conditional form of the Lebesgue dominated convergence theorem. The only problem connected with this is that of obtaining a conditional form of the monotone convergence theorem.

Theorem 7 (Conditional Form of Lebesgue Monotone Convergence Theorem). Let $\{X_n\}$ be a nondecreasing sequence of nonnegative random variables such that $X_n \to X$ a.s. as $n \to \infty$ and

$$X \in L_1(\Omega, \mathcal{C}, P).$$

Then, for any sub-sigma field $\mathcal{B} \subset \mathcal{C}$,

$$E(X_n \mid \mathcal{B}) \to E(X \mid \mathcal{B}) \text{ a.s.} \qquad \text{as} \quad n \to \infty.$$

Proof: By Remark 4 above,

$$E(X_n \mid \mathcal{B}) \leq E(X_{n+1} \mid \mathcal{B}) \leq E(X \mid \mathcal{B}) \text{ a.s.} \qquad \text{for all } n.$$

Hence $\lim_{n \to \infty} E(X_n \mid \mathcal{B})$ exists almost surely and is $\mathcal{B}$-measurable, and thus all we need to prove is that, for all $B \in \mathcal{B}$,

$$\int_B (\lim_{n \to \infty} E(X_n \mid \mathcal{B})) \, dP = \int_B E(X \mid \mathcal{B}) \, dP.$$

Applying the monotone convergence theorem twice we obtain

$$\int_B (\lim_{n \to \infty} E(X_n \mid \mathcal{B})) \, dP = \lim_{n \to \infty} \int_B E(X_n \mid \mathcal{B}) \, dP$$

$$= \lim_{n \to \infty} \int_B X_n \, dP = \int_B X \, dP = \int_B E(X \mid \mathcal{B}) \, dP,$$

which proves the theorem.

Theorem 8 (Conditional Form of Fatou's Lemma). If $\{X_n\}$ is a sequence of nonnegative random variables with finite expectations, if $\mathfrak{B}$ is a sub-sigma field of $\mathfrak{A}$, and if $E(\liminf_{n\to\infty} X_n) < \infty$, then

$$E(\liminf_{n\to\infty} X_n \mid \mathfrak{B}) \leq \liminf_{n\to\infty} E(X_n \mid \mathfrak{B}) \text{ a.s.}$$

Proof: This theorem follows from Theorem 7 in the same way that Fatou's lemma is obtained from the monotone convergence theorem.

Theorem 9 (Conditional Form of the Lebesgue Dominated Convergence Theorem). If Y, X_1, X_2, $\cdots$ are random variables, if $|X_n| \leq Y$ a.s. for all n, if $Y \in L_1(\Omega, \mathfrak{A}, P)$, if $X_n \to X$ a.s., and if $\mathfrak{B}$ is a sub-sigma field of $\mathfrak{A}$, then

$$E(X_n \mid \mathfrak{B}) \to E(X \mid \mathfrak{B}) \text{ a.s.} \qquad \text{as} \quad n \to \infty.$$

Proof: This theorem follows from Theorem 8 in the same way that the ordinary dominated convergence theorem follows from Fatou's lemma.

In the subsequent treatment of martingales and submartingales, a conditional form of Jensen's inequality will be needed. It will be recalled that a function g defined over $(-\infty, +\infty)$ is said to be a *convex function* if for every x and y and $\theta \in (0, 1)$ the inequality

$$g(\theta x + (1 - \theta)y) \leq \theta g(x) + (1 - \theta)g(y)$$

holds. Convex functions possess certain properties which we state here but do not prove. One property is that every convex function is continuous. Another property is that a convex function g is either monotone over $(-\infty, +\infty)$, or there is a finite number x_0 such that g is nonincreasing over $(-\infty, x_0]$ and nondecreasing over $[x_0, \infty)$. (See Exercise 5.) Last, there is Jensen's inequality which states that if X is a random variable, if g is a convex function, if $E(X)$ and $E(g(X))$ exist, then $g(E(X)) \leq E\,g(X)$.

Theorem 10 (Conditional Form of Jensen's Inequality). Let g be a convex function over $(-\infty, +\infty)$, and let X be a random variable such that X and $g(X)$ have finite expectations. If $\mathfrak{B}$ is any sub-sigma field of $\mathfrak{A}$, then

$$g(E(X \mid \mathfrak{B})) \leq E(g(X) \mid \mathfrak{B}) \text{ a.s.}$$

Proof: We need only prove the theorem in the case that there exists an x_0 such that g is nonincreasing over $(-\infty, x_0]$ and nondecreasing over $[x_0, \infty)$. (This covers the other two cases anyway when we take $x_0 = -\infty$ and $x_0 = +\infty$.) Without loss of generality we may assume that $x_0 = 0$ and $g(x_0) = 0$. We first prove the theorem in the case that X is discrete and takes only a finite set of values, say, $X = \sum_{n=1}^N b_n I_{B_n}$, where $\cup_{n=1}^N B_n = \Omega$ and $\{B_1, \cdots, B_N\}$ are disjoint events. First we observe that $\sum_{n=1}^N E(I_{B_n} \mid \mathfrak{B}) = 1$ a.s. and $E(I_{B_n} \mid \mathfrak{B}) \geq 0$ a.s. for $1 \leq n \leq N$. Then we obtain

$$E(g(X) \mid \mathfrak{B}) = E(g(\sum_{n=1}^N b_n I_{B_n}) \mid \mathfrak{B})$$

$$= E(\sum_{n=1}^N g(b_n) I_{B_n} \mid \mathfrak{B}) = \sum_{n=1}^N g(b_n) E(I_{B_n} \mid \mathfrak{B})$$

$$\geq g(\sum_{n=1}^N b_n E(I_{B_n} \mid \mathfrak{B})) = g(E(X \mid \mathfrak{B})).$$

Now let us define

$$X_n = \sum_{k=1}^{n2^n} \left(\frac{k-1}{2^n}\right) I_{[(k-1)2^{-n} \leq X < k2^{-n}]}$$

$$+ \sum_{k=-2^n n}^{-1} \left(\frac{k+1}{2^n}\right) I_{[(k+1)2^{-n} \geq X > k2^{-n}]}.$$

Clearly, X_n is discrete and takes only a finite number of values. In addition, $|X_n| \leq |X|$, $X_n \to X$ a.s. as $n \to \infty$, $|X_n| \leq n$, and, most important,

$$0 \leq g(X_n) \leq g(X) \in L_1(\Omega, \mathfrak{A}, P).$$

By what we proved above,

$$g(E(X_n \mid \mathfrak{B})) \leq E(g(X_n) \mid \mathfrak{B}) \text{ a.s.}$$

By Theorem 9, $E(X_n \mid \mathfrak{B}) \to E(X \mid \mathfrak{B})$ a.s., and since g is continuous we have $g(E(X_n \mid \mathfrak{B})) \to g(E(X \mid \mathfrak{B}))$ a.s. Since $0 \leq g(X_n) \leq g(X)$, and since $g(X_n) \to g(X)$ a.s., we apply Theorem 9 again to obtain $E(g(X_n) \mid \mathfrak{B}) \to E(g(X) \mid \mathfrak{B})$ a.s. This establishes the inequality. Q.E.D.

We conclude this section with a brief exposition of an alternative approach to conditional expectation. Let $X \in L_1(\Omega, \mathfrak{A}, P)$ and let Y

be any random variable. For every Borel set $B \subset (-\infty, +\infty)$, let us define a set function φ by

$$\varphi(B) = \int_{[Y \in B]} X \, dP.$$

It is clear that φ is a finite signed measure over the sigma field of Borel subsets of $(-\infty, +\infty)$. Let us define the measure μ_Y by

$$\mu_Y(B) = P[Y \in B]$$

for every Borel set B. Clearly φ is absolutely continuous with respect to μ_Y, and the Radon-Nikodym theorem applies. We define $E(X \mid Y = y)$ to be a Borel-measurable function uniquely determined except over a Borel set of μ_Y-measure zero by

$$\varphi(B) = \int_B E(X \mid Y = y) \, d\mu_Y(y).$$

In other words, $E(X \mid Y = y)$ is a Borel-measurable function of y and is the Radon-Nikodym derivative of φ with respect to μ_Y, that is,

$$E(X \mid Y = y) = \frac{d\varphi}{d\mu_Y}(y).$$

We now show the connection between this definition of conditional expectation and the previous one.

Theorem 11. Let $X \in L_1(\Omega, \mathfrak{a}, P)$ and let Y be any random variable. If Φ is a function defined by $\Phi(y) = E(X \mid Y = y)$, then $\Phi(Y) = E(X \mid Y)$ a.s.

Proof: Since by hypothesis Φ is Borel-measurable, and because of Theorem 1, we have that both $\Phi(Y)$ and $E(X \mid Y)$ are $\sigma\{Y\}$-measurable. Hence we need only prove

$$\int_B \Phi(Y) \, dP = \int_B E(X \mid Y) \, dP$$

for all $B \in \sigma\{Y\}$. Because of the definition of $E(X \mid Y)$ we need only prove that

$$\int_B \Phi(Y) \, dP = \int_B X \, dP \qquad \text{for all} \quad B \in \sigma\{Y\}.$$

For $B \in \sigma\{Y\}$ there is a Borel set $C \subset (-\infty, +\infty)$ such that

$$B = [Y \in C].$$

Hence

$$\int_B \Phi(Y)\, dP = \int_{[Y \in C]} \Phi(Y)\, dP = \int_C \Phi(y)\, dF_Y(y) = \int_C \Phi(y)\, d\mu_Y(y).$$

Letting φ be as defined above, then

$$\Phi(y) = \frac{d\varphi}{d\mu_Y}(y).$$

Hence

$$\int_B \Phi(Y)\, dP = \int_C \frac{d\varphi}{d\mu_Y}(y)\, d\mu_Y(y) = \int_C d\varphi(y) = \varphi(C)$$

$$= \int_{[Y \in C]} X\, dP = \int_B X\, dP.$$

Q.E.D.

The principal theorems of this section are Theorems 3, 4, and 10.

EXERCISES

1. Let $X \in L_1(\Omega, \mathcal{A}, P)$, let $\mathcal{B}$ be a sub-sigma field of $\mathcal{A}$, and define

$$\varphi(B) = \int_B X\, dP \qquad \text{for all} \quad B \in \mathcal{B}.$$

Prove that φ is a finite signed measure and that φ is absolutely continuous with respect to P over $(\Omega, \mathcal{B})$.

2. Prove Remarks 4 and 5.

3. If X and Y are in $L_1(\Omega, \mathcal{B}, P)$, prove that $X = Y$ a.s. if and only if

$$\int_B X\, dP = \int_B Y\, dP \qquad \text{for all} \quad B \in \mathcal{B}.$$

4. Prove that every convex function is continuous.

5. Prove that if g is a convex function over $(-\infty, +\infty)$, then g is monotone or there is an x_0 such that g is nonincreasing over $(-\infty, x_0]$ and nondecreasing over $[x_0, \infty)$.

6. Let $(\Omega, \, \alpha, \, P)$ be the unit interval probability space, let $\mathfrak{B}$ be the sub-sigma field generated by $\{[0, \frac{1}{4}], (\frac{1}{4}, \frac{2}{3}], (\frac{2}{3}, 1]\}$, and let $X(\omega) = \omega^2$ for all $\omega \in \Omega = [0, 1]$. Prove that $E(X \mid \mathfrak{B})$ can be expressed as

$$E(X \mid \mathfrak{B}) \;=\; a_1 I_{[0,1/4]} \,+\, a_2 I_{(1/4,2/3]} \,+\, a_3 I_{(2/3,1]},$$

and compute a_1, a_2, a_3.

7. If Y is a discrete random variable, if $P[Y = y_n] > 0$ for $n = 1, 2, \cdots$, and $\sum_n P[Y = y_n] = 1$, and if, for $X \in L_1(\Omega, \, \alpha, \, P)$,

$$\varphi(y) \;=\; \begin{cases} \displaystyle\int X \, dP(\cdot \mid [Y = y_n]) & \text{if} \quad y = y_n \\[2em] 0 & \text{if} \quad y \notin \{y_n\}, \end{cases}$$

then $\varphi(y) = E(X \mid Y = y)$ a.e. $[\mu_Y]$.

8. Prove: If $X \in L_1(\Omega, \, \alpha, \, P)$, if $\mathfrak{B}$ is a sub-sigma field of α, and if $\sigma\{X\}$ and $\mathfrak{B}$ are two independent classes of events, then $E(X \mid \mathfrak{B}) = EX$ a.s.

7.2. Martingales and Submartingales

One of the most fortunate circumstances that can occur when one wishes to prove almost sure convergence or convergence in rth mean of a particular sequence of random variables is that of being able to recognize the sequence as a martingale or submartingale. These next two sections are devoted to the basic properties and convergence theorems in martingale theory, and they are followed by a section devoted to an application of martingale theory. This section is concerned with properties of martingales and submartingales.

Definition: Let $\mathfrak{X} = \{X_\tau, \, \tau \in T\}$ be a collection of random variables, indexed by a simply ordered set $\{T, \, \leq\}$. $\mathfrak{X}$ is called a martingale if $E \mid X_\tau \mid < \infty$ for all $\tau \in T$ and if $X_s = E(X_t \mid \sigma\{X_\tau, \, \tau \leq s\})$ a.s. for $s < t$, $s \in T$ and $t \in T$.

The order types of T that are frequently encountered are: (a) any set of real numbers under natural ordering, for example, $[0, \infty)$, the positive integers, or the negative integers; (b) the set of all finite ordinals together with the first transfinite ordinal; and (c) one element *followed by* all negative integers in their natural ordering. If T is the set of all positive integers $\{n\}$ under their natural ordering, then $\{X_n\}$ is called a martingale sequence. The best way to remember the definition is to remember that of a fair game; in this case, X_n denotes a player's fortune upon completion of the nth play, and a fair game is determined by the fact that if one knows one's fortunes $X_1, \cdots, X_n$, then his expected fortune upon completion of the $(m + n)$th play is his present fortune X_n, that is, $X_n = E(X_{n+m} \mid X_1, \cdots, X_n)$.

Remark 1. If $\mathfrak{X} = \{X_\tau, \tau \in T\}$ is a martingale, then EX_τ does not depend on τ. This follows by taking expectations of both sides in the above definition.

Definition: A collection of random variables $\mathfrak{X} = \{X_\tau, \tau \in T\}$ is called a submartingale if T is a simply ordered set with ordering relation $\leq$, if $E \mid X_\tau \mid < \infty$ for all $\tau \in T$ and if $X_s \leq E(X_t \mid \sigma\{X_\tau, \tau \leq s\})$ a.s. for $s < t$.

The easiest way to remember the definition of a submartingale is to consider it as a record of fortunes of a player of a game in which the game is either fair or is favorable to the player.

Remark 2. If $\mathfrak{X} = \{X_\tau, \tau \in T\}$ is a submartingale, then $EX_s \leq EX_t$ for $s < t, s \in T, t \in T$.

Definition: If $\mathfrak{X} = \{X_t, t \in T\}$ is a collection of random variables, if T is simply ordered, then $\mathfrak{X}$ is called a reverse martingale if it is a martingale under the reverse ordering of T, and $\mathfrak{X}$ is called a reverse submartingale if it is a submartingale under the reverse ordering of T.

From the definition, then, a sequence of random variables $\{X_n\}$ is a reverse martingale if $X_{m+n} = E(X_n \mid X_{m+n}, X_{m+n+1}, \cdots)$ a.s., and it is a reverse submartingale if

$$X_{m+n} \leq E(X_n \mid X_{m+n}, X_{m+n+1}, \cdots) \text{ a.s.} \quad \text{for} \quad m \geq 1, n \geq 1.$$

Example 1. Let $\{X_n\}$ be a sequence of independent random variables with $EX_n = 0$ for all n, and let $S_n = X_1 + \cdots + X_n$. Then one can show that $\{S_n\}$ is a martingale. Indeed, since $\sigma\{S_1, \cdots, S_n\} = \sigma\{X_1, \cdots, X_n\}$, we have

$$E(S_{n+m} \mid S_1, \cdots, S_n) = E(S_{n+m} \mid X_1, \cdots, X_n)$$

$$= E(X_{n+1} + \cdots + X_{n+m} \mid X_1, \cdots, X_n) + E(S_n \mid X_1, \cdots, X_n)$$

$$= 0 + S_n,$$

using Theorems 4 and 5 of Section 7.1.

Example 2. Let $Y \in L_1(\Omega, \alpha, P)$, let $\{X_n\}$ be any sequence of random variables, and define $Z_n = E(Y \mid X_n, X_{n+1}, \cdots)$. We show that $\{Z_n\}$ is a reverse martingale by use of Theorems 3 and 4 of Section 7.1. Indeed, since

$$\sigma\{X_n, X_{n+1}, \cdots\} \supset \sigma\{Z_n, Z_{+1}, \cdots\}$$

(which follows from Theorem 1 in Section 7.1) we have by Theorem 3 in Section 7.1

$$E(Z_n \mid Z_{n+m}, Z_{n+m+1}, \cdots)$$

$$= E(E(Z_n \mid X_{n+m}, X_{n+m+1}, \cdots) \mid Z_{n+m}, Z_{n+m+1}, \cdots).$$

Again by Theorem 3 of Section 7.1,

$$E(Z_n \mid X_{n+m}, X_{n+m+1}, \cdots)$$

$$= E(E(Y \mid X_n, X_{n+1}, \cdots) \mid X_{n+m}, X_{n+m+1}, \cdots)$$

$$= E(Y \mid X_{n+m}, X_{n+m+1}, \cdots) = Z_{n+m} \text{ a.s.}$$

Thus from the above two equations, and by either Theorem 4 in Section 7.1 or by the uniqueness of conditional expectation we obtain

$$E(Z_n \mid Z_{n+m}, Z_{n+m+1}, \cdots) = E(Z_{n+m} \mid Z_{n+m}, Z_{n+m+1}, \cdots) = Z_{n+m} \text{ a.s.},$$

which proves the assertion.

Notation: If x is a real number we define $x^+ = \max\{x, 0\}$ and $x^- = \max\{-x, 0\} = -\min\{x, 0\}$. If X is a random variable, then $X^+ = XI_{[X \geq 0]}$ and $X^- = -XI_{[X \leq 0]}$. Clearly $x = x^+ - x^-$ and $X = X^+ - X^-$.

Theorem 1. Let $\{X_n\}$ be a submartingale. Then $\{X_n^+\}$ is a submartingale.

Proof: Let g be a real-valued function defined over $(-\infty, +\infty)$ by $g(x) = x^+$. Then g is a nondecreasing, convex function. By Theorem 1 of Section 1.3, $\sigma\{X_1^+, \cdots, X_n^+\} \subset \sigma\{X_1, \cdots, X_n\}$. Hence by Theorems 3 and 10 in Section 7.1, and because of the fact that g is monotone, we have

$$E(X^+_{n+m} \mid X_1^+, \cdots, X_n^+) = E(g(X_{n+m}) \mid X_1^+, \cdots, X_n^+)$$

$$= E(E(g(X_{n+m}) \mid X_1, \cdots, X_n) \mid X_1^+, \cdots, X_n^+)$$

$$\geq E(g(E(X_{n+m} \mid X_1, \cdots, X_n)) \mid X_1^+, \cdots, X_n^+)$$

$$\geq E(g(X_n) \mid X_1^+, \cdots, X_n^+)$$

$$= E(X_n^+ \mid X_1^+, \cdots, X_n^+) = X_n^+ \text{ a.s.} \quad \text{Q.E.D.}$$

Theorem 2. If $\{X_n\}$ is a martingale, then $\{|X_n|\}$ is a submartingale.

Proof: Let g be a function defined over $(-\infty, +\infty)$ by $g(x) = |x|$. Then g is convex. By Theorem 1 of Section 1.3 we obtain

$$\sigma\{|X_1|, |X_2|, \cdots, |X_n|\} \subset \sigma\{X_1, \cdots, X_n\}.$$

Thus by Theorem 10 of Section 7.1 we have

$$|X_n| = g(X_n) = g(E(X_{n+m} \mid X_1, \cdots, X_n))$$

$$\leq E(g(X_{n+m}) \mid X_1, \cdots, X_n) = E(|X_{n+m}| \mid X_1, \cdots, X_n) \text{ a.s.}$$

Now we take the conditional expectation of both sides given $\sigma\{|X_1|, \cdots, |X_n|\}$ and obtain

$$|X_n| \leq E(|X_{n+m}| \mid |X_1|, \cdots, |X_n|).$$

Q.E.D.

The above two theorems establish that certain functions of martingales and/or submartingales are submartingales. The next two theorems yield inequalities which are used to establish uniform integrability of martingales in the next section.

Theorem 3. If $X_1, X_2, \cdots, X$ is a martingale (note the order type), then for every $\epsilon > 0$,

$$P[\sup_n |X_n| > \epsilon] \leq \frac{1}{\epsilon} \int_{[\sup_n |X_n| > \epsilon]} |X| \, dP.$$

Proof: By Theorem 2, $|X_1|, |X_2|, \cdots, |X|$ is a submartingale, and in particular

$$|X_j| \leq E(|X| \mid |X_1|, \cdots, |X_j|) \quad \text{a.s.}$$

for $j = 1, 2, \cdots$. Let us denote

$$A_j = [|X_j| > \epsilon] \cap \bigcap_{i=1}^{j-1} [|X_i| \leq \epsilon] \quad \text{and} \quad B = [\sup_n |X_n| > \epsilon].$$

The events $\{A_n\}$ are disjoint, and, further,

$$A_n \in \sigma\{|X_1|, \cdots, |X_n|\} \quad \text{and} \quad B = \bigcup_{j=1}^{\infty} A_j.$$

By the definition of conditional expectation we have

$$\int_B |X| \, dP = \sum_{n=1}^{\infty} \int_{A_n} |X| \, dP = \sum_{n=1}^{\infty} \int_{A_n} E(|X| \mid |X_1|, \cdots, |X_n|) \, dP$$

$$\geq \sum_{n=1}^{\infty} \int_{A_n} |X_n| \, dP \geq \epsilon \sum_{n=1}^{\infty} P(A_n) = \epsilon P(B),$$

which proves the inequality.

Theorem 4. If $\{X_n\}$ is a reverse martingale, then, for every $\epsilon > 0$,

$$P[\sup_n |X_n| > \epsilon] \leq \epsilon^{-1} \int_{[\sup_n |X_n| > \epsilon]} |X_1| \, dP.$$

Proof: Since $X_n, X_{n-1}, \cdots, X_1$ is a martingale (in this order), then by the same proof as that used in Theorem 3 we obtain

$$P(B_n) \leq \epsilon^{-1} \int_{B_n} |X_1| \, dP,$$

where $B_n = [\max_{1 \leq k \leq n} |X_k| > \epsilon]$. Let $n \to \infty$ on both sides of the above inequality, and since

$$B_n = \bigcup_{j=1}^{n} [|X_j| > \epsilon] \to \bigcup_{j=1}^{\infty} [|X_j| > \epsilon] = [\sup_n |X_n| > \epsilon],$$

we obtain the conclusion.

The remainder of this section is devoted to a proof of the crucial inequality in Theorem 5 which in turn is used in the next section to prove the martingale convergence theorem.

Definition: Let $\{x_n\}$ be a sequence of real numbers, and let $-\infty < a < b < \infty$. Define $k_1 = \min \{i \mid x_i \leq a\}$,

$$k_2 = \min \{i \mid i > k_1, x_i \geq b\},$$
$$\vdots$$
$$k_{2j+1} = \min \{i \mid i > k_{2j}, x_i \leq a\},$$
$$k_{2j+2} = \min \{i \mid i > k_{2j+1}, x_i \geq b\}.$$

Define $h_n = \max \{j \mid k_{2j} \leq n\}$. Then h_n is called the number of upcrossings of $[a, b]$ by $x_1, \cdots, x_n$.

Notation. For $k = 3, 4, \cdots$, we define $\{i_k\}$ as follows:

$$i_k = \begin{cases} 1 & \text{if } k_{2j} < k \leq k_{2j+1}, \text{ for some } j = 1, 2, \cdots \\ 0 & \text{otherwise.} \end{cases}$$

The following lemma clarifies the meaning of $\{i_k\}$.

Lemma 1. (a) $i_k = 1$ if and only if for some $j < k$, $x_j \leq a$ and for some $r, j < r < k$, $x_r \geq b$ and $x_{r+1} > a, \cdots, x_{k-1} > a$, and (b) $i_k = 0$ if and only if for some $j < k$, $x_j \leq a$, $x_{j+1} < b, \cdots, x_{k-1} < b$ or if $x_1 > a, \cdots$, $x_{k-1} > a$. In other words, i_k depends on x_j for $j < k$.

Proof: This lemma is easily seen to be equivalent to the above definition of $\{i_k\}$.

The important point about Lemma 1 is the last sentence.

Lemma 2.

$$\sum_{r=3}^{n} i_r(x_r - x_{r-1}) \leq (a - b)h_n + (x_n - a)^+.$$

Proof: If $h_n = 0$, then $i_r = 0$ for $1 \leq r \leq n$, and in this case the lemma is true. Consider next the case where $h_n > 0$, $k_{2h_n} \leq n$ and $k_{2h_n+1} > n$.

In this case,

$$\sum_{r=3}^{n} i_r(x_r - x_{r-1}) = (x_{k_3} - x_{k_2}) + (x_{k_5} - x_{k_4}) + \cdots$$

$$+ (x_{k_{2h_n-1}} - x_{k_{2h_n-2}}) + (x_n - x_{k_{2h_n}})$$

$$\leq (a - b)(h_n - 1) + (x_n - a) + (a - x_{k_{2h_n}}).$$

But $a - x_{k_{2h_n}} \leq a - b$ and $x_n - a > 0$. So

$$\sum_{r=3}^{n} i_r(x_r - x_{r-1}) \leq (a - b)h_n + (x_n - a) = (a - b)h_n + (x - a)^+.$$

The last case to consider is the case where

$$h_n > 0, \; k_{2h_n+1} \leq n \quad \text{and} \quad k_{2h_n+2} > n.$$

Then

$$\sum_{r=3}^{n} i_r(x_r - x_{r-1}) \leq (x_{k_3} - x_{k_2}) + \cdots + (x_{k_{2h_n+1}} - x_{k_{2h_n}}) \leq (a - b)h_n,$$

from which the lemma also follows. Q.E.D.

Let $\{X_n\}$ be a sequence of random variables. In connection with such a sequence we define the extended real valued $\mathcal{C}$-measurable functions $\{K_n\}$ over Ω by

$$K_1 = \min \{i \mid X_i \leq a\} = \sum_{n=1}^{\infty} n I_{[X_n \leq a]} \bigcap_{j=1}^{n-1} [X_j > a] + \infty \cdot I_{\bigcap_{n=1}^{\infty} [X_n > a]},$$

$$K_2 = \min \{i \mid i > K_1, X_i \geq b\},$$

and in general,

$$K_{2m+1} = \min \{i \mid i > K_{2m}, X_i \leq a\}$$

and

$$K_{2m} = \min \{i \mid i > K_{2m+1}, X_i \geq b\}.$$

Upon observing the expression for K_1 it is easy to verify that every K_n is $\mathcal{C}$-measurable. Next let us define

$$H_n = \max \{j \mid K_{2j} \leq n\} = \sum_{m=1}^{\infty} m I_{[K_{2m} \leq n][K_{2m+2} > n]}.$$

Clearly $\{H_n\}$ is a nondecreasing sequence of nonnegative random variables, and $H_n \leq [n/2]$. Finally, for $k = 3, 4, \cdots$, let I_k denote the indicator of the event

$$\bigcup_{j=1}^{k-2} \{[X_j \leq a] \cap (\bigcup_{r=j+1}^{k-1} [X_r \geq b][X_{r+1} > a] \cdots [X_{k-1} > a])\}.$$

Lemma 3. The sequences $\{I_n\}$ and $\{H_n\}$ obey the following properties:

(i) $\displaystyle\sum_{r=3}^{n} I_r(X_r - X_{r-1}) \leq (a - b)H_n + (X_n - a)^+$, and

(ii) $\sigma\{I_k\} \subset \sigma\{X_1, X_2, \cdots, X_{k-1}\}$.

Proof: This lemma is a restatement of Lemma 2 and its proof is the same.

Theorem 5. If $\{X_n\}$ is a submartingale, and if a, b, H_n are as above, then $(b - a)E(H_n) \leq E(X_n - a)^+$. If $\{X_n\}$ is a reverse submartingale, and if H'_n denotes the number of upcrossings of the ordered finite sequence $X_n, X_{n-1}, \cdots, X_1$ over $[a, b]$, then $(b - a)EH'_n \leq E(X_1 - a)^+$.

Proof: We first consider the case when $\{X_n\}$ is a submartingale. By Lemma 3,

$$\sum_{k=3}^{n} I_k(X_k - X_{k-1}) \leq (a - b)H_n + (X_n - a)^+.$$

Since $\sigma\{I_k\} \subset \sigma\{X_1, \cdots, X_{k-1}\}$, since $I_k \geq 0$, and since $E(X_k \mid X_1, \cdots, X_{k-1}) - X_{k-1} \geq 0$ a.s., we have by Theorem 4 of Section 7.1

$$E(I_k(X_k - X_{k-1})) = E\{E(I_k(X_k - X_{k-1}) \mid X_1, \cdots, X_{k-1})\}$$

$$= E\{I_k E(X_k - X_{k-1} \mid X_1, \cdots, X_{k-1})\}$$

$$= E\{I_k(E(X_k \mid X_1, \cdots, X_{k-1}) - X_{k-1})\} \geq 0$$

and is finite. Since $0 \leq H_n \leq n/2$, then $EH_n < \infty$, and thus, taking expectations of both sides in Lemma 3 we get

$$(b - a)EH_n \leq E(X_n - a)^+.$$

The same proof goes through in the case of a reverse submartingale but with different subscripts, the last line of the proof being

$$(b - a)EH'_n \leq E(X_1 - a)^+.$$

Q.E.D.

EXERCISES

1. Let $\{X_n\}$ be a sequence of independent random variables with corresponding characteristic functions $\{f_n(u)\}$. Assume there is an interval I such that $f_n(u) \neq 0$ for all n and all $u \in I$. For each $u \in I$, define

$$Y_n(u) = \frac{\exp\{iu \sum_{j=1}^{n} X_j\}}{\prod_{j=1}^{n} f_j(u)}.$$

Prove that for each fixed $u \in I$, the sequence $\{Y_n(u)\}$ is a martingale.

2. Prove: If $\{X_n\}$ is any sequence of random variables, if

$$S_n = X_1 + \cdots + X_n,$$

then $\sigma\{X_1, \cdots, X_n\} = \sigma\{S_1, \cdots, S_n\}$.

3. Let (Ω, α, P) be the unit-interval probability space. For every n, let $\theta_{n,1}, \theta_{n,2}, \cdots, \theta_{n,2^{n-1}}$ be 2^{n-1} numbers in $(0, 1)$. Let Π_1 be the set of intervals $\{[0, \theta_{1,1}), [\theta_{1,1}, 1)\}$, and if $[a_j, b_j)$ is the jth contiguous interval from the left in Π_{n-1}, let Π_n consist of intervals $[a_j, a_j + \theta_{nj}(b_j - a_j))$, $[a_j + \theta_{nj}(b_j - a_j), b_j)$. Let $a_{n,1}, \cdots, a_{n,2^n}$ be any real numbers for every n. Define

$$X_1 = 2I_{[0,\theta_{1,1})} + 3I_{[\theta_{1,1},1)}.$$

Assuming X_{n-1} defined and constant over each of the 2^{n-1} disjoint intervals in Π_{n-1}, define X_n by

$$X_n(\omega) = \begin{cases} X_{n-1}(\omega) + (1 - \theta_{nj})a_{n-1,j} & \text{if } \omega \in [a_j, a_j + \theta_{nj}(b_j - a_j)) \\ X_{n-1}(\omega) - \theta_{nj}a_{n-1,j} & \text{if } \omega \in [a_j + \theta_{nj}(b_j - a_j), b_j). \end{cases}$$

Prove that $\{X_n\}$ is a martingale.

7.3. Martingale and Submartingale Convergence Theorems

The machinery so far developed in this chapter is applied in this section to prove the basic martingale and submartingale convergence theorems. Theorem 1 appears to be the most widely used theorem in the research literature and is generally the one referred to as "the martingale convergence theorem."

Theorem 1. If $\{X_n\}$ is a submartingale, and if $\sup_n E \mid X_n \mid \, < \, \infty$, then there is a random variable X such that $X_n \to X$ a.s. and $E \mid X \mid \, \leq \sup_n E \mid X_n \mid$.

Proof: Let A denote the event that the sequence $\{X_n\}$ does not converge, and let

$$A_{r,s} = [\liminf X_n \leq r < s \leq \limsup X_n],$$

where $r < s$. It is easy to verify that

$$A = \cup\{A_{r,s} \mid r < s, r \text{ and } s \text{ are rational}\}.$$

In order to prove that there exists an X such that $X_n \to X$ a.s. we need only prove that $PA_{r,s} = 0$ if $r < s$. Let $-\infty < r < s < \infty$, and let H_n denote the number of upcrossings over $[r, s]$ by $X_1, \cdots, X_n$. Now $0 \leq H_n \leq H_{n+1}$ a.s. for all n, and thus there exists an H (which is possibly infinite over an event of positive probability) such that $H_n \to H$ a.s. By the Lebesgue *monotone* convergence theorem and by Theorem 5 in Section 7.2 we have

$$\int H \, dP = \sup_n EH_n \leq \frac{\sup_n E(X_n - r)^+}{s - r}.$$

But

$$E(X_n - r)^+ \leq E \mid X_n - r \mid \, \leq E \mid X_n \mid + \mid r \mid.$$

By hypothesis we have

$$\int H \, dP \leq \frac{\sup_n (E \mid X_n \mid + \mid r \mid)}{s - r} < \infty,$$

which in turn implies that H is finite a.s. One easily verifies that

$$A_{r,s} \subset [H = \infty],$$

and since $P[H = \infty] = 0$, then $PA_{r,s} = 0$, which proves there is an X such that $X_n \to X$ a.s. In addition, $|X_n| \to |X|$ a.s., so by Fatou's lemma and by hypothesis

$$E|X| \leq \liminf_n E|X_n| \leq \sup E|X_n| < \infty,$$

which concludes the proof of the theorem.

For reverse submartingales a better theorem (in a sense) can be obtained, namely, that *every* reverse submartingale converges almost surely, although the limiting $\mathcal{C}$-measurable function need not necessarily have a finite integral. For reverse martingales with no additional hypotheses one gets not only convergence almost surely but convergence in rth mean ($r = 1$). These two results will be obtained in Theorem 2 and the corollary to Theorem 3.

Theorem 2. If $\{X_n\}$ is a reverse submartingale, then there is an $\mathcal{C}$-measurable function X such that $X_n \to X$ a.s. (Note that X is not called a random variable; this is done because X might be infinite with positive probability.)

Proof: Let A and $A_{r,s}$ be as defined in the proof of Theorem 1, let $r < s$ be finite numbers, and let H'_n denote the number of upcrossings over $[r, s]$ by $X_n, X_{n-1}, \cdots, X_2, X_1$ (note the order). Again, it is easy to see that H'_n is nonnegative, finite, and nondecreasing. By the second part of Theorem 5 in Section 7.2 we have, for every n,

$$E|H'_n| \leq \frac{E(X_1 - r)^+}{(s - r)} \leq \frac{E|X_1| + |r|}{s - r}.$$

Since there is an $\mathcal{C}$-measurable function H' over Ω such that $H'_n \to H'$ a.s., we again have, by the monotone convergence theorem, that

$$\int H' \, dP = \sup EH'_n \leq \frac{E|X_1| + r}{s - r}$$

which is finite. Hence H' is finite almost surely, and by the same argument as that used in the proof of Theorem 1 we obtain $P(A_{r,s}) = 0$. Thus $X_n \to X$ a.s., where X is some $\mathcal{C}$-measurable function over Ω. Q.E.D.

In order to obtain stronger convergence theorems it is necessary to introduce the notion of uniform integrability.

Definition: A collection of random variables $\{X_\alpha, \ \alpha \in A\}$ is said to be uniformly integrable if for every $\epsilon > 0$ there is a number $K_\epsilon > 0$ such that $E(\ |\ X_\alpha\ |\ I_{[|X_\alpha| \geq K_\epsilon]}) < \epsilon$ for all $\alpha \in A$.

The following two lemmas give basic properties of uniform integrability needed here.

Lemma 1. If a collection of random variables $\{X_\alpha, \ \alpha \in A\}$ is uniformly integrable, then for every $\epsilon > 0$ there exists a $\delta_\epsilon > 0$ such that if $P(B) < \delta_\epsilon$ then $E(\ |\ X_\alpha\ |\ I_B) < \epsilon$ for all $\alpha \in A$.

Proof: For any event B and every $\alpha \in A$,

$$E(\ |\ X_\alpha\ |\ I_B) = E(\ |\ X_\alpha\ |\ I_{B \cap [|X_\alpha| \leq K]})$$
$$+ E(\ |\ X_\alpha\ |\ I_{B \cap [|X_\alpha| > K]})$$
$$\leq KP(B) + E(\ |\ X_\alpha\ |\ I_{[|X_\alpha| > K]}).$$

For $\epsilon > 0$, select $K > 0$ so large that $E(\ |\ X_\alpha\ |\ I_{[|X_\alpha| > K]}) < \epsilon/2$ for all $\alpha \in A$. For this value of K select $\delta_\epsilon = \epsilon/2K$. Hence for all $B \in \mathcal{Q}$ for which $P(B) < \delta_\epsilon$, the inequality $E(\ |\ X_\alpha\ |\ I_B) < \epsilon$ holds for all $\alpha \in A$. Q.E.D.

Lemma 2. If $\{X_n\}$ is a sequence of random variables with finite expectations, if $X_n \to X$ a.s., where X is a random variable, and if $\{X_n\}$ are uniformly integrable, then $X_n \to X$ in rth mean, where $r = 1$.

Proof: Since $\{X_n\}$ are uniformly integrable, then by Lemma 1, for every $\epsilon > 0$ there is a $\delta_\epsilon > 0$ such that if $P(B) < \delta_\epsilon$, then $E(\ |\ X_n\ |\ I_B) < \epsilon$ for all n. Since $X_n \to X$ a.s., then by Egorov's theorem there is an event A such that $P(A) < \delta_\epsilon$ and such that $X_n \to X$ uniformly over A^c. Hence

$$E\ |\ X_n - X_m\ | \leq E(\ |\ X_n - X_m\ |\ I_{A^c}) + E(\ |\ X_n\ |\ I_A) + E(\ |\ X_m\ |\ I_A)$$
$$\leq E(\ |\ X_n - X_m\ |\ I_{A^c}) + 2\epsilon.$$

Since $X_n \to X$ uniformly over A^c, we have

$$0 \leq E\ |\ X_n - X_m\ | < 3\epsilon$$

for n, m sufficiently large. But $L_1(\Omega, \mathcal{Q}, P)$ is complete, and hence there exists a random variable $X' \in L_1(\Omega, \mathcal{Q}, P)$ such that $X_n \to X'$ in rth mean, where $r = 1$. From Exercise 3 in Section 4.4 there is a subsequence $\{X_{n_k}\}$ of $\{X_n\}$ such that $X_{n_k} \to X'$ a.s. But $X_n \to X$ a.s., hence $X_{n_k} \to X$ a.s., so $X = X'$. Thus $X_n \to X$ in rth mean, where $r = 1$.

Theorem 3. If $\{X_n\}$ is a submartingale or a reverse submartingale, and if the random variables $\{X_n\}$ are uniformly integrable, then there is a random variable X such that $X_n \to X$ a.s. and in rth mean, where $r = 1$.

Proof: We first consider the case where $\{X_n\}$ is a submartingale. By the definition above, the hypothesis that $\{X_n\}$ are uniformly integrable means that for every $\epsilon > 0$ there is a finite number $K_\epsilon > 0$ such that $E(\,|X_n|\,I_{[|X_n| \geq K_\epsilon]}) < \epsilon$ for all n. Hence $E\,|X_n| \leq K_\epsilon + \epsilon$ for all n, which implies that

$$\sup_n E\,|X_n| \leq K_\epsilon + \epsilon < \infty.$$

By Theorem 1 there is a random variable X such that $X_n \to X$ a.s. and $E\,|X| \leq \sup_n E\,|X_n|$. Applying Lemma 2 we obtain that $X_n \to X$ in rth mean, where $r = 1$. If $\{X_n\}$ is a reverse submartingale, then by Theorem 2 there is an $\mathcal{Q}$-measurable function X such that $X_n \to X$ a.s. We now show that X is a random variable, that is, that X is finite with probability one. Let $A = [X = +\infty]$, and suppose that $P(A) = \alpha > 0$. Then, since $X_n \to \infty$ as $n \to \infty$ over A, we have $1/X_n \to 0$ over A. Hence by Egorov's theorem, $1/X_n \to 0$ uniformly over A except over an event of probability less than $\alpha/2$. Thus, given any $K > 0$ there is an N such that for all $n > N$, $X_n > K$ over an event of probability greater than $\alpha/2$, or $E(\,|X_n|\,I_{[|X_n| \geq K]}) > K\alpha/2$ for all $n > N$. This violation of uniform integrability gives us the contradiction sought and therefore guarantees the finiteness of X. Now by Lemma 2, $X_n \to X$ in rth mean $(r = 1)$. Q.E.D.

The following corollary shows that for a reverse martingale no further properties must be satisfied in order to obtain convergence almost surely and convergence in rth mean $(r = 1)$.

Corollary to Theorem 3. Let $\{X_n\}$ be a reverse martingale. Then there exists a random variable X such that $X_n \to X$ a.s. and in rth mean $(r = 1)$.

Proof: Let $B_n = [\,|X_n| > K\,]$ and

$$B = \bigcup_{n=1}^{\infty} B_n = [\sup_n |X_n| > K].$$

By Theorem 2 in Section 7.2, if $\{X_n\}$ is a reverse martingale, then $\{\,|X_n|\,\}$ is a reverse submartingale; that is,

$$|X_n| \leq E(\,|X_{n-k}|\,\|\,|X_n|,\,|X_{n+1}|,\,\cdots) \text{ a.s.} \qquad \text{for } 1 \leq k \leq n-1,$$

$n = 2, 3, \cdots$. But $B_n \in \sigma\{\,|X_n|,\,|X_{n+1}|,\,\cdots\}$ and $B_n \subset B$ for all n, so

$$\int_{B_n} |X_n|\,dP \leq \int_{B_n} E(\,|X_1|\,\|\,|X_n|,\,|X_{n+1}|,\,\cdots)\,dP$$

$$= \int_{B_n} |X_1|\,dP \leq \int_B |X_1|\,dP.$$

By Theorem 4 of Section 7.2,

$$P(B) \leq \frac{1}{K}\int_B |X_1|\,dP \leq \frac{E\,|X_1|}{K} \to 0 \qquad \text{as } K \to \infty.$$

Hence for $\epsilon > 0$ there is a $K > 0$ such that

$$\int_B |X_1|\,dP < \epsilon,$$

which by the above inequality implies that

$$\int_{B_n} |X_n|\,dP < \epsilon$$

for all n, that is, $\{X_n\}$ are uniformly integrable. The conclusion of the corollary follows immediately from Theorem 3.

Definition: Let $\{X_n\}$ be a martingale. Then X is called a closing random variable for $\{X_n\}$ if $X_1, X_2, \cdots, X$ is a martingale. We say that X is a nearest closing random variable for $\{X_n\}$ if for any other closing random variable Y, then $X_1, X_2, \cdots, X, Y$ is a martingale.

Theorem 4. Let $X_1, X_2, \cdots, Y$ be a martingale. Then there is a random variable $X \in L_1(\Omega, \mathcal{G}, P)$ such that $X_n \to X$ a.s. and in rth mean $(r = 1)$, and X is a nearest closing random variable of $\{X_n\}$.

Proof: Let us denote

$$B_n = [\,|\,X_n\,| > K\,] \text{ and}$$

$$B = \bigcup_{n=1}^{\infty} B_n = [\sup |\,X_n\,| > K].$$

We may apply Theorem 3 in Section 7.2 to obtain

$$P[\sup |\,X_n\,| > K] \le \frac{1}{K} \int_B |\,Y\,|\,dP \le \frac{E\,|\,Y\,|}{K}.$$

Since $E\,|\,Y\,| < \infty$, this last inequality implies that

$$P[\sup_n |\,X_n\,| > K] \to 0 \qquad \text{as} \quad K \to \infty.$$

By Theorem 2 in Section 7.2, $|\,X_1\,|, |\,X_2\,|, \cdots, |\,Y\,|$ is a submartingale. Thus

$$|\,X_n\,| \le E(\,|\,Y\,|\,|\,|\,X_1\,|, |\,X_2\,|, \cdots, |\,X_n\,|\,) \text{ a.s.,}$$

and since $B_n \in \sigma\{\,|\,X_1\,|, \cdots, |\,X_n\,|\}$, we obtain

$$\int_{B_n} |\,X_n\,|\,dP \le \int_{B_n} |\,Y\,|\,dP.$$

From the fact that $B_n \subset B$ it follows that

$$\int_{B_n} |\,X_n\,|\,dP = E(\,|\,X_n\,|\,I_{B_n}) \le \int_B |\,Y\,|\,dP$$

for all n. For K sufficiently large, $P(B)$ becomes as small as we choose (this was established above), and hence for any $\epsilon > 0$ there is a $K > 0$ large enough so that

$$\int_B |\,Y\,|\,dP < \epsilon.$$

Thus $\{X_n\}$ are uniformly integrable. By Theorem 3 there is a random variable X such that $X_n \to X$ a.s. and in rth mean ($r = 1$). We next prove that X is a closing random variable for $\{X_n\}$. Let $C_n \in \sigma\{X_1, \cdots, X_n\}$. Since $X_n = E(X_{n+m} \mid X_1, \cdots, X_n)$ a.s. we first note that

$$\int_{C_n} X_n\,dP = \int_{C_n} E(X_{n+m} \mid X_1, \cdots, X_n)\,dP = \int_{C_n} X_{n+m}\,dP.$$

Secondly, $X_k \to X$ in rth mean $(r = 1)$ as $k \to \infty$, that is,

$$E \mid X_k - X \mid \to 0 \qquad \text{as} \quad k \to \infty,$$

which implies that for fixed n,

$$0 \le E(\mid X_k - X \mid I_{C_n}) \le E \mid X_k - X \mid \to 0 \qquad \text{as} \quad k \to \infty.$$

From this latter observation we obtain

$$\left| \int_{C_n} X_{n+m} \, dP - \int_{C_n} X \, dP \right| \le E(\mid X_{n+m} - X \mid I_{C_n}) \to 0 \quad \text{as } m \to \infty,$$

or

$$\int_{C_n} X_{n+m} \, dP \to \int_{C_n} X \, dP \qquad \text{as} \quad m \to \infty.$$

Thus, from our first observation above we obtain

$$\int_{C_n} X_n \, dP = \int_{C_n} X \, dP$$

for all n and all $C_n \in \sigma\{X_1, \cdots, X_n\}$. By the uniqueness of conditional expectation, we obtain $X_n = E(X \mid X_1, \cdots, X_n)$ a.s., that is, X_1, $X_2, \cdots, X$ is a martingale, and X is a closing random variable for $\{X_n\}$. Finally we prove that X is a nearest closing random variable of $\{X_n\}$. Let Z be any closing random variable of $\{X_n\}$, that is,

$$X_n = E(Z \mid X_1, \cdots, X_n) \text{ a.s.}$$

for every n. Let $C_n \in \sigma\{X_1, \cdots, X_n\}$. As before,

$$\int_{C_n} X_n \, dP = \int_{C_n} E(Z \mid X_1, \cdots, X_n) \, dP = \int_{C_n} Z \, dP$$

or

$$\int_{C_n} X_n \, dP = \int_{C_n} Z \, dP.$$

Now let $C \in \sigma\{X_1, X_2, \cdots\} = \sigma\{X_1, X_2, \cdots, X\}$. For every n there is a $C_n \in \sigma\{X_1, \cdots, X_n\}$ such that $P(C_n \bigtriangleup C) = P(C_n C^c \cup C_n{}^c C) \to 0$

as $n \to \infty$. Next we observe that

$$\left| \int_{C_n} X_n \, dP - \int_C X \, dP \right| \leq \left| \int_{C_n} X_n \, dP - \int_C X_n \, dP \right|$$

$$+ \left| \int_C X_n \, dP - \int_C X \, dP \right|$$

$$\leq \int |X_n| I_{C_n \Delta C} \, dP + \int_C |X_n - X| \, dP$$

$$\leq E(|X_n| I_{C_n \Delta C}) + E|X_n - X|.$$

We have already proved that $\{X_n\}$ are uniformly integrable, and thus by Lemma 1 and for sufficiently large n, $E(|X_n| I_{C_n \Delta C})$ becomes arbitrarily small. In addition, $E|X_n - X| \to 0$ as $n \to \infty$, and thus

$$\int_{C_n} X_n \, dP \to \int_C X \, dP \qquad \text{as} \quad n \to \infty.$$

One easily verifies that

$$\int_{C_n} Z \, dP \to \int_C Z \, dP.$$

It was shown above that

$$\int_{C_n} Z \, dP = \int_{C_n} X_n \, dP,$$

from which we obtain

$$\int_{C_n} X_n \, dP \to \int_C Z \, dP.$$

Thus

$$\int_C X \, dP = \int_C Z \, dP \qquad \text{for all} \quad C \in \sigma\{X_1, X_2, \cdots, X\};$$

that is, $X = E(Z \mid X_1, X_2, \cdots, X)$ a.s. or $X_1, X_2, \cdots, X, Y$ is a martingale. Q.E.D.

The following is a very particular martingale theorem.

Theorem 5. Let $Y \in L_1(\Omega, \mathcal{C}, P)$, and let $\{X_n\}$ be a sequence of random variables. Then $E(Y \mid X_1, \cdots, X_n) \to E(Y \mid X_1, X_2, \cdots)$ a.s. and in rth mean $(r = 1)$ as $n \to \infty$.

Proof: Let us denote $Z_n = E(Y \mid X_1, \cdots, X_n)$ and

$$Z = E(Y \mid X_1, X_2, \cdots).$$

We first prove that $Z_1, Z_2, \cdots, Z$ is a martingale. By Theorem 1 in Section 7.1 one observes that

$$\sigma\{Z_1, \cdots, Z_n\} \subset \sigma\{X_1, \cdots, X_n\},$$

from which we obtain by Theorems 3 and 4 in Section 7.1 that

$$E(Z_{n+1} \mid Z_1, \cdots, Z_n) = E(E(Z_{n+1} \mid X_1, \cdots, X_n) \mid Z_1, \cdots, Z_n)$$

$$= E(Z_n \mid Z_1, \cdots, Z_n) = Z_n \text{ a.s.}$$

Similarly $E(Z \mid Z_1, \cdots, Z_n) = Z_n$ a.s. By Theorem 4, since $\{Z_n\}$ is a martingale closed by Z, there exists a random variable Z_0 such that $Z_n \to Z_0$ a.s. and in rth mean $(r = 1)$. We must therefore prove that $Z_0 = Z$ a.s. Let $B \in \mathrm{U}_{n=1}^{\infty} \sigma\{X_1, \cdots, X_n\}$. Then there is a positive integer m such that $B \in \sigma\{X_1, \cdots, X_m\}$. It is easy to verify that

$$\left| \int_B Z_0 \, dP - \int_B Z \, dP \right|$$

$$\leq \left| \int_B Z_0 \, dP - \int_B Z_k \, dP \right| + \left| \int_B Z_k \, dP - \int_B Z \, dP \right|$$

$$\leq E \mid Z_0 - Z_k \mid + \left| \int_B Z_k \, dP - \int_B Z \, dP \right|.$$

For $k \geq m$,

$$\int_B Z_k \, dP = \int_B Y \, dP = \int_B Z \, dP,$$

and also $E \mid Z_0 - Z_k \mid \to 0$ as $k \to \infty$. Thus

$$\int_B Z_0 \, dP = \int_B Z \, dP \quad \text{for all} \quad B \in \bigcup_{n=1}^{\infty} \sigma\{X_1, \cdots, X_n\}.$$

Thus by the extension theorem for measures we have

$$\int_B Z_0 \, dP = \int_B Z \, dP \quad \text{for all} \quad B \in \sigma\{X_1, X_2, \cdots\}.$$

Since both Z_0 and Z are measurable with respect to $\sigma\{X_1, X_2, \cdots\}$ it follows that $Z_0 = Z$ a.s. Q.E.D.

Corollary to Theorem 5. A sequence of random variables $\{X_n\}$ is a reverse martingale if and only if

$$E(X_n \mid X_{n+k}, X_{n+k+1}, \cdots, X_{n+k+m}) = X_{n+k}$$

a.s. for every n, $k \geq 0$, $m \geq 0$.

Proof: If $\{X_n\}$ is a reverse martingale, then by definition

$$E(X_n \mid X_{n+k}, X_{n+k+1}, \cdots) = X_{n+k} \text{ a.s.}$$

If we take the conditional expectation of both sides given the sigma field $\sigma\{X_{n+k}, \cdots, X_{n+k+m}\}$, we obtain the equation given in the corollary. Conversely, given the equation in the statement of the corollary, we know by Theorem 5 that

$$E(X_n \mid X_{n+k}, X_{n+k+1}, \cdots, X_{n+k+m}) \to E(X_n \mid X_{n+k}, X_{n+k+1}, \cdots) \text{ a.s.}$$

as $m \to \infty$. Thus the defining equality for a martingale holds. Q.E.D.

EXERCISES

1. Let $\{X_n\}$ be a sequence of random variables. Prove that the event that $\{X_n\}$ does not converge is the same as

$$\bigcup_{r<s} [\liminf X_n \leq r < s \leq \limsup X_n],$$

where the union is taken over all pairs of rationals r, s such that $r < s$.
2. Construct an example of a finite sequence of real numbers $x_1, x_2, \cdots, x_n$ and an interval $[a, b]$ such that the number of upcrossings of x_1, $x_2, \cdots, x_n$ over $[a, b]$ is not the same as the number of upcrossings of $x_n, x_{n-1}, \cdots, x_1$ over $[a, b]$.
3. In the notation of the proof of Theorem 1, prove that

$$A_{r,s} \subset [H = \infty].$$

4. Prove: If $\{X_n\}$ is a martingale, then there is at most one nearest closing random variable for the sequence.

5. Prove: If $\{X_n\}$ is a sequence of random variables, and if $X_n \to X$ a.s., then $\sigma\{X_1, X_2, \cdots\} = \sigma\{X_1, X_2, \cdots, X\}$.

6. Prove: If $X_n \to X$ as $n \to \infty$ in rth mean $(r = 1)$, then $\{X_1, X_2, \cdots, X\}$ are uniformly integrable.

7. If $\{x_n\}$ is a sequence of real numbers, and if h'_n denotes the number of upcrossings of $x_n, x_{n-1}, \cdots, x_2, x_1$ (note the order) over the interval $[a, b]$, prove that h'_n is nondecreasing.

8. Let $\mathfrak{X} = \{X_n, n = 0, \pm 1, \pm 2, \cdots\}$ be an ordered set of random variables. Prove that $\mathfrak{X}$ is a submartingale if and only if

$$X_n \leq E(X_{n+m} \mid X_r, \cdots, X_n) \text{ a.s.} \qquad \text{for all} \quad r, n, m, r \leq n, m \geq 0.$$

7.4. Brownian Motion

In this section an application of the martingale convergence theorem is given. This application is a celebrated theorem due to Paul Lévy on Brownian motion, and its inclusion in this chapter serves as an introduction to the chapter on stochastic processes.

Definition: Let $\mathfrak{X} = \{X(t), t \in [0, \infty)\}$ be a collection of random variables. (That is, for each $t \in [0, \infty)$, $X(t)$ is a random variable which assigns to $\omega \in \Omega$ the number $X(t, \omega)$.). $\mathfrak{X}$ is called a Brownian motion (or a Brownian motion stochastic process) if it satisfies the following three requirements:

(a) $\mathfrak{X}$ has independent increments; that is, if n is any positive integer and if $0 = t_0 < t_1 < \cdots < t_n < \infty$, then the n random variables

$$\{X(t_j) - X(t_{j-1}), 1 \leq j \leq n\}$$

are independent,

(b) the distribution function of $X(t) - X(s)$ is $\mathfrak{N}(0, (t - s)\sigma^2)$, where $0 \leq s < t < \infty$ and $\sigma^2 > 0$ does not depend on s or t, and

(c) $P[X(0) = 0] = 1$. The collection $\mathfrak{X}$ is usually denoted by $X(t)$.

The first problem considered is that of showing that a Brownian motion stochastic process does exist.

Theorem 1. There exists a probability space $(\Omega, \mathcal{C}, P)$ and a collection of random variables $\mathfrak{X} = \{X(t), 0 \leq t < \infty)$ over it which is a Brownian motion for a given $\sigma^2 > 0$.

Proof: Let n be any positive integer, and let $\{t_0, t_1, \cdots, t_n\}$ be any $n + 1$ real numbers which satisfy $0 = t_0 < t_1 < \cdots < t_n < \infty$. Let $X_1, X_2, \cdots, X_n$ be n independent random variables (defined over some probability space $(\Omega', \mathcal{C}', P')$) such that $\mathcal{L}(X_j) = \mathfrak{N}(0, \sigma^2(t_j - t_{j-1}))$, $1 \leq j \leq n$. For every n and every such $\{t_1, \cdots, t_n\}$ let us define the function

$$F_{t_0}(x) = 0 \quad \text{if} \quad x < 0 \quad \text{and} \quad F_{t_0}(x) = 1 \quad \text{if} \quad x \geq 0,$$

$$F_{t_1, \cdots, t_n}(x_1, \cdots, x_n) = P(\bigcap_{j=1}^{n} [S_j \leq x_j]),$$

where $S_j = X_1 + \cdots + X_j$, $1 \leq j \leq n$. Let $\mathfrak{F}$ denote the set of all functions $\{F_{t_1, \cdots, t_n}\}$ obtained in this way for all positive integers n and all selections $\{t_0, t_1, \cdots, t_n\}$ which satisfy $0 = t_0 < t_1 < \cdots < t_n < \infty$. The family of functions $\mathfrak{F}$ satisfies the hypotheses of the Kolmogorov-Daniell theorem (Theorem 2 in Section 2.3), and hence there exists by that theorem a probability space $(\Omega, \mathcal{C}, P)$ and a set of random variables $\mathfrak{X} = \{X(t), t \in [0, \infty)\}$ such that

$$F_{t_1, \cdots, t_n}(x_1, \cdots, x_n) = P(\bigcap_{j=1}^{n} [X(t_j) \leq x_j])$$

for all $F_{t_1, \cdots, t_n} \in \mathfrak{F}$. Trivially, condition (c) in the definition is satisfied. In order to verify condition (a) in the definition, let $Y_j = X(t_j) - X(t_{j-1})$, $1 \leq j \leq n$, let $X_1, \cdots, X_n$ be n independent random variables (defined over some other probability space $(\Omega', \mathcal{C}', P')$) such that $\mathcal{L}(X_j) = \mathfrak{N}(0, (t_j - t_{j-1})\sigma^2)$, $1 \leq j \leq n$, and let $f_{t_1, \cdots, t_n}$ denote the joint characteristic function of $X(t_1), \cdots, X(t_n)$. Then, for $(y_1, \cdots, y_n) \in E^{(n)}$

$$E(\exp i \sum_{j=1}^{n} y_j Y_j)$$

$$= E(\exp \{i((y_1 - y_2)X(t_1) + (y_2 - y_3)X(t_2) + \cdots$$
$$+ (y_{n-1} - y_n)X(t_{n-1}) + y_n X(t_n))\})$$

$$= f_{t_1, \cdots, t_n}(y_1 - y_2, y_2 - y_3, \cdots, y_{n-1} - y_n, y_n)$$

$$= E(\exp \{i((y_1 - y_2)X_1 + (y_2 - y_3)(X_1 + X_2) + \cdots$$
$$+ (y_{n-1} - y_n)(X_1 + \cdots + X_{n-1}) + y_n(X_1 + \cdots + X_n))\})$$

$$= E \exp \{i(y_1 X_1 + y_2 X_2 + \cdots + y_n X_n)\}.$$

Since $X_1, \cdots, X_n$ are independent we have

$$E \exp \left(i \sum_{j=1}^{n} y_j Y_j \right) = \prod_{j=1}^{n} E \exp \left(i y_j X_j \right).$$

But

$$E(\exp i y_j X_j) = E(\exp \{ i(y_j(X_1 + \cdots + X_j) - y_j(X_1 + \cdots + X_{j-1})) \})$$

$$= f_{t_1, \cdots, t_n}(0, \cdots, 0, -y_j, y_j, 0, \cdots, 0),$$

where $-y_j$ occurs in the $(j-1)$th place. Thus

$$E(\exp i y_j X_j) = f_{t_{j-1}, t_j}(-y_j, y_j) = f_{Y_j}(y_j).$$

Hence by Theorem 2 of Section 3.2, $Y_1, Y_2, \cdots, Y_n$ are independent, which proves (a). In order to verify (b) in the definition we note that

$$f_{Y_j}(y_j) = f_{t_{j-1}, t_j}(-y_j, y_j) = f_{s_{j-1}, s_j}(-y_j, y_j) = f_{X_j}(y_j).$$

Hence the distribution of $X(t_j) - X(t_{j-1})$ is $\mathfrak{N}(0, (t_j - t_{j-1})\sigma^2)$, which verifies (b). Q.E.D.

In order to obtain the theorem desired we need one computation and one lemma.

Computation: If X is a random variable whose distribution is $\mathfrak{N}(0, \sigma^2)$, then Var $(X^2) = 2\sigma^4$.

Verification. The distribution of X/σ is $\mathfrak{N}(0, 1)$ and $EX^4 = \sigma^4 E(X/\sigma)^4$. But

$$E(X/\sigma)^4 = (1/\sqrt{2\pi}) \int_{-\infty}^{\infty} x^4 \exp (-x^2/2) \, dx$$

$$= \sqrt{2/\pi} \int_{0}^{\infty} x^4 \exp (-x^2/2) \, dx$$

$$= (4/\sqrt{\pi}) \int_{0}^{\infty} u^{3/2} e^{-u} \, du$$

$$= 4\Gamma(5/2)/\sqrt{\pi} = 3.$$

Hence Var $(X^2) = E(X^4) - (E(X^2))^2 = 3\sigma^4 - \sigma^4 = 2\sigma^4$.

Lemma 1. If X is a random variable with finite expectation and symmetric distribution, if $\mathcal{B}$ is a sub-sigma field of $\mathcal{A}$, and if $\mathcal{B}$ and $\sigma\{X\}$ are independent, then $E(X \mid \mathcal{C}) = 0$ a.s., where $\mathcal{C}$ is the sigma field generated by $\mathcal{B}$ and $\sigma\{X^2\}$.

Proof: Let $0 \le a < b \le \infty$, and let $B \in \mathcal{B}$. Then

$$B \cap [a < X^2 \le b] \in \mathcal{C}.$$

Let us denote

$$K = \int_{B \cap [a < X^2 \le b]} E(X \mid \mathcal{C}) \, dP.$$

Then

$$K = \int I_B I_{[a < X^2 \le b]} X \, dP.$$

But I_B and $XI_{[a < X^2 \le b]}$ are independent. Hence

$$K = P(B) \int_{[a < X^2 \le b]} X \, dP$$

$$= P(B) \left(\int_{-\sqrt{b}-0}^{-\sqrt{a}-0} + \int_{\sqrt{a}+0}^{\sqrt{b}+0} x \, dF(x) \right),$$

from which we obtain $K = 0$ by the symmetry of $F(x) = P[X \le x]$. In the same way one obtains

$$\int_{B \cap [X^2 \in L]} E(X \mid \mathcal{C}) \, dP = 0,$$

where L is any Borel-measurable subset of $[0, \infty)$. Since the same equation holds when integrating over finite disjoint unions of events of the form $B \cap [X^2 \in L]$ (which are a field which generates $\mathcal{C}$), we have by the uniqueness of the extension theorem that

$$\int_C E(X \mid \mathcal{C}) \, dP = 0 \qquad \text{for all} \quad C \in \mathcal{C},$$

which implies the conclusion of the lemma.

We now have all the machinery needed to prove the principal theorem of this section.

Theorem 2 (P. Lévy). Let $X(t)$, $0 \le t < \infty$, be a Brownian motion stochastic process with Var $(X(t)) = \sigma^2 t$ and $EX(t) = 0$. Let

$$\mathcal{P}_n = \{t_{n,0}, t_{n,1}, \cdots, t_{n,n}\}$$

be such that $0 = t_{n,0} < t_{n,1} < \cdots < t_{n,n} = T$ for all n and $T > 0$ fixed, $\mathcal{P}_n \subset \mathcal{P}_{n+1}$ for all n, and max $\{t_{n,j} - t_{n,j-1} \mid 1 \le j \le n\} \to 0$ as $n \to \infty$. Then

$$\sum_{j=1}^{n} (X(t_{n,j}) - X(t_{n,j-1}))^2 \to \sigma^2 T \text{ a.s.}$$

and in rth mean $(r = 1)$.

Proof: Let us denote

$$S_n = \sum_{j=1}^{n} (X(t_{n,j}) - X(t_{n,j-1}))^2.$$

We shall prove that the sequence of random variables $\{S_n\}$ is a reverse martingale. By the corollary to Theorem 5 in Section 7.3 we need only prove that $E(S_n \mid S_{n+1}, \cdots, S_{n+m}) = S_{n+1}$ a.s. for all n, $m \ge 1$, or, equivalently, that

(1) $$E(S_n - S_{n+1} \mid S_{n+1}, \cdots, S_{n+m}) = 0 \text{ a.s.}$$

We shall prove (1) in a case where $m = 2$ and shall leave it to the reader to see that this proof holds in general. Since $\mathcal{P}_n \subset \mathcal{P}_{n+1}$ for all n, and since $X(t)$ has independent increments, we need only prove the following statement: if X_1, X_2, X_3, S are independent random variables, and if $\mathcal{L}(X_i) = \mathfrak{N}(0, \sigma_i^2)$, $i = 1, 2, 3$, then

(2) $$E((X_1 + X_2 + X_3)^2 + S - (X_1 + X_2)^2 - X_3^2 - S \mid (X_1 + X_2)^2$$
$$+ X_3^2 + S, X_1^2 + X_2^2 + X_3^2 + S)$$
$$= 2E((X_1 + X_2)X_3 \mid X_1^2 + X_2^2 + X_3^2$$
$$+ S, (X_1 + X_2)^2 + X_3^2 + S) = 0.$$

In order to prove this we note that

$$E((X_1 + X_2)X_3 \mid X_1, X_2, X_3^2, S) = (X_1 + X_2)E(X_3 \mid X_1, X_2, X_3^2, S) \text{ a.s.}$$

In Lemma 1, let $\mathcal{B} = \sigma\{X_1, X_2, S\}$. Now, applying Lemma 1 we see that $E(X_3 \mid X_1, X_2, X_3^2, S) = 0$ a.s., or

(3) $$E((X_1 + X_2)X_3 \mid X_1, X_2, X_3^2, S) = 0 \text{ a.s.}$$

Now take the conditional expectation of both sides of (3) given the sigma field $\sigma\{X_1^2 + X_2^2 + X_3^2 + S, (X_1 + X_2)^2 + X_3^2 + S\}$, and we obtain (2). The proof for the cases $m = 3, 4, \cdots$ follows the same ideas as above. Hence $\{S_n\}$ is a reverse martingale, and thus by the corollary to Theorem 3 in Section 7.3 we know that there exists a random variable Z such that $S_n \to Z$ a.s. and in rth mean ($r = 1$). Yet to be proved is the fact that $Z = \sigma^2 T$ a.s. It is easy to verify by known properties of variance and by the computation made prior to Lemma 1 that $ES_n = \sigma^2 T$ and

$$E(S_n - \sigma^2 T)^2 = \operatorname{Var} S_n$$

$$= \sum_{j=1}^{n} \operatorname{Var} (X(t_{n,j}) - X(t_{n,j-1}))^2$$

$$= \sum_{j=1}^{n} 2\sigma^4 (t_{n,j} - t_{n,j-1})^2$$

$$\leq 2\sigma^4 T \max \{t_{n,j} - t_{n,j-1} \mid 1 \leq j \leq n\} \to 0 \qquad \text{as } n \to \infty.$$

Thus by Chebishev's inequality, $S_n \xrightarrow{P} \sigma^2 T$. But above we showed $S_n \to Z$ a.s. which implies $S_n \xrightarrow{P} Z$. By the uniqueness of limit in probability we obtain $Z = \sigma^2 T$ a.s. Q.E.D.

EXERCISES

1. Prove that if X and Y are independent random variables, if X has finite expectation, and if the distribution of X is symmetric, then $E(X \mid X^2 + Y^2) = 0$ a.s.
2. If $X(t)$ is a Brownian motion, evaluate $\operatorname{Cov} (X(t), X(t + h))$ for $h > 0$.
3. In the probability space $(\Omega, \mathcal{C}, P)$ constructed in Theorem 1, the probability P really depends on σ; denote it by P_σ. Prove that $P_{\sigma_1} \perp P_{\sigma_2}$ if and only if $\sigma_1 \neq \sigma_2$. (Hint: use Theorem 2.) (**Definition:** two probability measures P_1 and P_2 are said to be orthogonal if there exist two disjoint events A_1 and A_2 such that $P_1(A_1) = P_2(A_2) = 1$, in which case we write $P_1 \perp P_2$.)

CHAPTER 8

An Introduction to Stochastic Processes and, in Particular, Brownian Motion

8.1. Probability Measures over Function Spaces

A stochastic process is simply a collection of random variables, $\{X_t, t \in T\}$, all defined over the same probability space, where T is an indexing set and X_t is a random variable for each t. Sometimes the random variable X_t will be denoted by $X(t)$, and the value this random variable assigns to the elementary event ω will be denoted by $X(t, \omega)$. The stochastic process $\{X_t, t \in T\}$ or $\{X(t), t \in T\}$ will sometimes be denoted either by X_t or $X(t)$. In this chapter a foundation for the theory of stochastic processes is established, and a particular stochastic process known as separable Brownian motion will be investigated to some depth. In this section some basic questions on the nature of stochastic processes are discussed.

We have already done a great deal of work on a very special stochastic process, namely, a sequence of random variables. In this particular case no problems arise concerning the structure of the space Ω, because all operations are countable. However, if T is uncountable and, as is very frequently the case, is an interval of real numbers, then problems of uncountable operations occur.

If $\{X_t, t \in T\}$ is a stochastic process defined over some probability space $(\Omega, \mathcal{C}, P)$, then from the traditional point of view (see discussion at the end of Section 2.3) one is interested only in those problems which can be treated by being given the set of all joint distribution functions

$$F_{t_1,\cdots,t_n}(x_1, \cdots, x_n) = P(\bigcap_{j=1}^{n} [X_{t_j} \leq x_j])$$

for all finite subsets $\{t_1, \cdots, t_n\} \subset T$. This set of functions $\{F_{t_1,\cdots,t_n}:$ $\{t_1, \cdots, t_n\} \subset T\}$ satisfies the hypotheses of the Kolmogorov-Daniell theorem (Theorem 2 in Section 2.3). Consequently by this theorem as it was proved in Section 2.3, if we take Ω' as the set of all real-valued functions defined over T, if $\mathcal{C}'$ is the sigma field of subsets of Ω' generated by

$$\{\{f \in \Omega' \mid f(t) \leq x\}, t \in T, -\infty < x < \infty\},$$

and if X'_t is defined over Ω' by $X'_t(f) = f(t)$ for all $t \in T$ and all $f \in \Omega'$, then there exists a probability measure P' over $\mathcal{C}'$ which satisfies

$$F_{t_1,\cdots,t_n}(x_1, \cdots, x_n) = P'(\bigcap_{j=1}^{n} \{f \in \Omega' \mid X'_{t_j}(f) \leq x_j\})$$

for all $(x_1, \cdots, x_n) \in E^{(n)}$ and every finite subset $\{t_1, \cdots, t_n\} \subset T$. In other words, from a traditional point of view one might as well consider X_t as a coordinate function over the space of all real-valued functions over T.

There are what appear to be two problems encountered in the foundations of the theory of stochastic processes. One problem is that of defining a probability of a set of elementary events obtained by, say, an uncountable intersection or union of events which one might very well wish to be an event. For example, if $\{X_t, t \in T\}$ is a stochastic process, one might wish to consider as an event the set

$$[X_t \leq x, t \in M] = \bigcap_{t \in M} [X_t \leq x],$$

where M is an uncountable subset of T and in particular is an interval. There *are* certain stochastic processes defined over certain probability spaces where such a set of elementary events is not an event. For example, let $(\Omega, \mathcal{C}, P)$ be the unit-interval probability space, let $T = [0, 1]$, and let S be a nonmeasurable subset of T. If $t \in S$, define

$$X_t(\omega) = \begin{cases} 1 & \text{if} \quad \omega = t \\ \\ 0 & \text{if} \quad \omega \neq t, \end{cases}$$

and if $t \notin S$, define $X_t(\omega) = 0$ for all $\omega \in \Omega$. Clearly X_t as defined is a stochastic process. Then

$$\{\omega \mid X_t(\omega) > \tfrac{1}{2} \text{ for some } t \in T\} = \bigcup_{t \in T} [X_t > \tfrac{1}{2}] = S,$$

or $[\sup_{t \in T} X_t \le \tfrac{1}{2}] = \bigcap_{t \in T}[X_t \le \tfrac{1}{2}] = [0, 1] \backslash S$, which is nonmeasurable since S is nonmeasurable. However, for every t,

$$P[X_t = 0] = P[X_t \le \tfrac{1}{2}] = 1,$$

and so it seems quite natural to wish to consider $[\sup_{t \in T} X_t \le \tfrac{1}{2}]$ as an event and to take its probability to be one.

Hence, given a stochastic process $\{X_t, t \in T\}$, we might rightfully wish to do the following in order to be able to define probabilities of sets of elementary events, say, of the form $\{\omega \mid \sup_{t \in S} X_t(\omega) \le x\}$, where S is an open interval. Namely, for every t, define a random variable X'_t so that the joint distribution of $X'_{t_1}, \cdots, X'_{t_n}$ is the same as that of $X_{t_1}, \cdots, X_{t_n}$ for all finite subsets $\{t_1, \cdots, t_n\} \subset T$ and such that every set of elementary events like that defined above, but now defined in terms of X'_t, is an event.

A second problem encountered in the theory of stochastic processes is in fitting the process over its proper function space. What this means is this: Let $\{X_t, t \in T\}$ be a stochastic process over a probability space $(\Omega, \mathcal{A}, P)$. It might occur that one wishes to take a particular set $\mathcal{F}_T$ of functions over T, the sigma field of subsets of $\mathcal{F}_T$, $\mathcal{B}$, generated by

$$\{\{f \in \mathcal{F}_T \mid f(t) \le x\}, t \in T, -\infty < x < \infty\},$$

and a probability measure P_0 defined over $\mathcal{B}$ which satisfies

$$P_0\{f \in \mathcal{F}_T \mid f(t_1) \le x_1, \cdots, f(t_n) \le x_n\} = P(\bigcap_{j=1}^{n} [X_{t_j} \le x_j])$$

for every $(x_1, \cdots, x_n) \in E^{(n)}$ and every finite subset $\{t_1, \cdots, t_n\} \subset T$. However, $\mathcal{F}_T$ might be so small that no set function satisfying the last equation above can be extended to a measure over $\mathcal{B}$. Thus the problem arises concerning conditions under which a particular function space could serve as a probability space for a certain stochastic process. As an example, if one considers Brownian motion already encountered in Section 7.4, one might wish to define it over the set of all continuous functions over $[0, \infty)$ which vanish at 0. It is not at all easy to see that this can be done, and only a rather complicated sequence of arguments can show that this can be done. This will be shown in Section 8.3.

As a simple example of whether a measure already defined on a space (say, the set of all functions) can be inherited by a well-spread-out subset (say, the set of all continuous functions), let us consider the unit interval probability space $(\Omega, \mathcal{Q}, P)$, and let Ω_r be the set of all rationals in Ω. Let $\mathcal{Q}_r$ be the sigma field of all subsets of Ω_r. For every interval $(a, b]$ with rational end points define $P_r((a, b] \cap \Omega_r) = b - a$; easily, P_r is defined over the field generated by $\{(a, b] \cap \Omega_r\}$ but is not completely additive over it.

The two problems discussed above are not too far apart, and both are solved by means of the notion of *separability* of a stochastic process, which we consider in the next section.

Before concluding this section some traditional terminology should be recorded. If one is dealing with just one stochastic process $\{X_t, t \in T\}$ over some probability space $(\Omega, \mathcal{Q}, P)$, then the points ω can indeed be represented by real-valued functions over T. We need only make correspond to each element ω in Ω the function f_ω defined by $f_\omega(t) = X_t(\omega)$ for all $t \in T$. Then we may replace ω by f_ω and define $X_t(f_\omega) = X_t(\omega)$. The functions f_ω will be referred to as *sample functions* of the process $\{X_t, t \in T\}$, and the expression "almost all sample functions" will mean "almost surely" or "with probability one."

Exercises

1. Let $T = [0, \infty)$, let Ω be the set of all *continuous* functions over T, let $\mathcal{Q}$ be the sigma field generated by

$$\{\{\omega \in \Omega \mid \omega(t) \le x\}, t \in T, -\infty < x < \infty\}.$$

Let P be a probability measure over $(\Omega, \mathcal{Q})$, and let $\{X_t, t \in T\}$ be defined by $X_t(\omega) = \omega(t)$. Prove that if $0 \le a < b < \infty$, then

$$\sup \{X(t) \mid a \le t \le b\}$$

is a random variable.

2. Let $\{Y_n\}$ be independent, identically distributed random variables, where

$$F_{Y_1}(y) = \begin{cases} 0 & \text{if } y \le 0 \\ 1 - e^{-y} & \text{if } y > 0. \end{cases}$$

Let

$$X_t = \max \{n \mid Y_1 + Y_2 + \cdots + Y_n \le t\}.$$

Prove that X_t is a random variable for each t. What is the natural function space that $\{X_t, t \in T = [0, \infty)\}$ should be defined over?
3. Prove that the stochastic process defined in Exercise 2 has independent increments.

8.2. Separable Stochastic Processes

The purpose of this section is to provide a definition of separability of stochastic processes and to prove the fundamental theorem about separability.

The first theorem of this section is an existence theorem needed to establish the definition of a separable stochastic process.

Theorem 1. Let $\{X_\lambda, \lambda \in \Lambda\}$ be any set of random variables defined over some probability space $(\Omega, \mathcal{C}, P)$. Then there is an extended-valued $\mathcal{C}$-measurable function X defined over Ω, uniquely determined except over an event of probability zero, which satisfies:

(a) $P[X_\lambda \le X] = 1$ for every $\lambda \in \Lambda$,
(b) there is a countable subset $\{\lambda_1, \lambda_2, \cdots\} \subset \Lambda$ such that

$$\sup \{X_{\lambda_n} \mid n = 1, 2, \cdots\} = X \text{ a.s.},$$

and

(c) if Y is any other extended-valued $\mathcal{C}$-measurable function satisfying $P[X_\lambda \le Y] = 1$ for all $\lambda \in \Lambda$, then $P[X \le Y] = 1$.

Proof: Uniqueness is a trivial consequence of (b). Let

$$\mathfrak{X} = \{\max \{X_{\lambda_1}, \cdots, X_{\lambda_n}\} \mid \text{all finite } \{\lambda_1, \cdots, \lambda_n\} \subset \Lambda\}.$$

If U and V are in $\mathfrak{X}$, then clearly $\max \{U, V\} \in \mathfrak{X}$; that is, $\mathfrak{X}$ is a lattice. Let us define

$$c = \sup \{E(\text{Arctan } U) \mid U \in \mathfrak{X}\}.$$

Since $-\pi/2 < c \le \pi/2$, there is a sequence $\{U_n\}$ in $\mathfrak{X}$ such that $U_n \le U_{n+1}$ a.s. and $E(\text{Arctan } U_n) \to c$ as $n \to \infty$. Let $U = \sup_n \text{Arctan } U_n$, and denote $X = \tan U$. Clearly U and therefore X are $\mathcal{C}$-measurable. We shall prove that X satisfies (a), (b), and (c) above. In order to prove (a), let us assume to the contrary that there exists a $\lambda \in \Lambda$ such that $P[X_\lambda > X] > 0$, or, equivalently, $P[\text{Arctan } X_\lambda > \text{Arctan } X] > 0$.

Now $E(\max \{\text{Arctan } X_\lambda, \text{Arctan } X\}) \leq \pi/2$. Let $Y_n = \max \{\text{Arctan } X_\lambda,$ $\text{Arctan } U_n\}$. Then $Y_n \leq Y_{n+1}$ a.s. and $Y_n \to Z_0$ a.s. as $n \to \infty$, where $Z_0 = \max \{\text{Arctan } X_\lambda, \text{Arctan } X\}$. But $\tan Y_n \in \mathfrak{X}$, $|Y_n| \leq \pi/2$, $|Z_0| \leq \pi/2$, and thus by the Lebesgue dominated convergence theorem, $EY_n \to EZ_0$. But $EZ_0 > E(\text{Arctan } X) = c$, and hence for sufficiently large n, $EY_n > c$, which contradicts the property of c. Hence

$$P[X_\lambda \leq X] = 1 \quad \text{for all} \quad \lambda \in \Lambda.$$

In order to prove (b) we simply note that $\text{Arctan } \max \{X_{\lambda_1}, \cdots, X_{\lambda_n}\} = \max \{\text{Arctan } X_{\lambda_1}, \cdots, \text{Arctan } X_{\lambda_n}\}$. Then since U_n in the proof of (a) can be written as $U_n = \max \{X_1, \cdots, X_{k_n}\}$ for a sequence $\{X_n\}$ in $\mathfrak{X}$ and for $k_n \to \infty$ as $n \to \infty$, and since $X = \lim_{n \to \infty} U_n$, we obtain $X = \sup_j X_j$, and $\{X_j\}$ is a countable subset of $\mathfrak{X}$ required. Condition (c) follows from (b). Indeed, if $P[X_\lambda \leq Y] = 1$ for all $\lambda \in \Lambda$, then it follows that

$$P[\sup_n X_{\lambda_n} \leq Y] = P(\overset{\infty}{\underset{n=1}{\cap}} [X_{\lambda_n} \leq Y]) = 1,$$

and since $\sup_n X_{\lambda_n} = X$ a.s., then (c) follows. Q.E.D.

Theorem 1 enables us to make the following definition.

Definition: If $\{X_\lambda, \lambda \in \Lambda\}$ is any collection of random variables over a probability space $(\Omega, \mathcal{A}, P)$, and if X is the unique extended-valued $\mathcal{A}$-measurable function whose existence is guaranteed by Theorem 1, then X is called the lattice-theoretic supremum of $\{X_\lambda, \lambda \in \Lambda\}$, and one denotes this by $X = l.\text{-th. sup } \{X_\lambda, \lambda \in \Lambda\}$. We define the lattice-theoretic infimum by $l.\text{-th. inf } \{X_\lambda, \lambda \in \Lambda\} = -l.\text{-th. sup } \{-X_\lambda, \lambda \in \Lambda\}$.

We may now define separable stochastic process.

Definition: Let $\mathfrak{X} = \{X_t, t \in T\}$ be a stochastic process defined over a probability space $(\Omega, \mathcal{A}, P)$, where T is an interval of real numbers. Then $\mathfrak{X}$ is called a separable stochastic process if for every open interval $(a, b) \subset T$,

$$l.\text{-th. sup } \{X_t, t \in (a, b)\} = \sup \{X_t, t \in (a, b)\} \text{ a.s.,}$$

and

$$l.\text{-th. inf } \{X_t, t \in (a, b)\} = \inf \{X_t, t \in (a, b)\} \text{ a.s.}$$

Definition: If $\{X_t, t \in T\}$ is a stochastic process defined over a probability space $(\Omega, \mathfrak{C}, P)$, and if $\{X'_t, t \in T\}$ is a stochastic process defined over the probability space $(\Omega', \mathfrak{C}', P')$, we shall say that these two processes are equivalent if the joint distribution functions of $X_{t_1}, \cdots, X_{t_n}$ and of $X'_{t_1}, \cdots, X'_{t_n}$ are the same for every finite subset $\{t_1, \cdots, t_n\}$ of T. (Note that T is the same in both cases.)

The following theorem is fundamental. It provides a solution to the first problem posed in Section 8.1.

Theorem 2. If $\{X_t, t \in T\}$ is any stochastic process defined over a probability space $(\Omega, \mathfrak{C}, P)$, and if T is an interval of real numbers, then there exists a separable stochastic process $\{X'_t, t \in T\}$, defined over the same probability space such that $P[X_t = X'_t] = 1$ for all $t \in T$, which in turn implies that the two processes are equivalent.

Proof: Let $\{(a_n, b_n)\}$ be the necessarily countable set of all open intervals such that a_n and b_n are rational and such that $(a_n, b_n) \subset T$. For every let

$$U_n = l.\text{-th. sup } \{X_\tau, \tau \in (a_n, b_n)\}$$

and

$$L_n = l.\text{-th. inf } \{X_\tau, \tau \in (a_n, b_n)\}.$$

Notice if $(a_r, b_r) = (a_n, b_n) \cap (a_m, b_m)$, then $U_r \le U_m$ a.s. and $L_r \ge L_m$. An easy argument about countable operations shows that one may change every U_n and L_n over events of probability zero so that they satisfy the definitions given to them above and in addition satisfy $U_n \ge U_r(\omega) \ge L_r(\omega) \ge L_m(\omega)$ for every $\omega \in \Omega$ if $(a_r, b_r) \subset (a_m, b_m)$ us denote, for each $t \in T$, $N_t = \{n \mid t \in (a_n, b_n)\}$, where t is not point of T. Clearly

$$\sup \{L_n(\omega) \mid n \in N_t\} \le \inf \{U_n(\omega) \mid n \in N_t\}$$

for all $t \in T$. Let us define

$$X_t^{(u)} = \min \{X_t, \inf \{U_n \mid n \in N_t\}\}.$$

Then X_t a.s., since $X_t \le U_n$ a.s. for all $n \in N_t$. Next let

$$X'_t = \max \{X_t^{(u)}, \sup \{L_n \mid n \in N_t\}\}.$$

For every $n \in N_t$, $L_n(\omega) \leq X'_t(\omega)$ for all $\omega \in \Omega$. In addition, $X_t^{(u)}(\omega) \leq U_n(\omega)$ for all $n \in N_t$ and all $\omega \in \Omega$, and hence by the definition of X'_t,

$$X'_t(\omega) \leq \max \{U_n(\omega), \sup \{L_m(\omega) \mid m \in I_t\}\}.$$

But

$$\sup \{L_m(\omega) \mid m \in N_t\} \leq \inf \{U_m(\omega) \mid m \in N_t\} \leq U_n(\omega)$$

for all $\omega \in \Omega$ and all $n \in N_t$, and hence $X'_t(\omega) \leq U_n)$; that is, we have

(1) $$L_n(\omega) \leq X'_t(\omega) \leq U_n(\omega)$$

for all $n \in N_t$ and all $\omega \in \Omega$. By the definition of X'_t, since $L_n \leq X_t$ a.s. and since $X_t^{(u)} = X_t$ a.s., we have that $X'_t = X_t$ a.s., that

$$P[X'_t = X_t] = 1 \quad \text{for every} \quad t \in T.$$

If t is an endpoint of T, define $X'_t = X_t$. Thus the two stochastic processes $\{X_t, t \in T\}$ and $\{X'_t, t \in T\}$ are both defined on the same probability space and are equivalent. We now prove that $\{X'_t \in T\}$ is separable. Let $(a, b) \subset T$; we must show that

$$l.\text{-th. sup } \{X'_t, t \in (a, b)\} = \sup \{X'_t, t \in (a, b)\} \text{ a.}$$

and

$$l.\text{-th. inf } \{X'_t, t \in (a, b)\} = \inf \{X'_t, t \in (a, b)\} \text{ a.s.}$$

We shall prove only the first relation, the proof of the second being similar to it. Let $\{(\alpha_n, \beta_n)\}$ be a sequence of intervals with rational endpoints such that $a < \alpha_{n+1} < \alpha_n < \beta_n < \beta_{n+1} < b$ for all n and that $\alpha_n \to a$ and $\beta_n \to b$ as $n \to \infty$. We clearly need only show that

$$l.\text{-th. sup } \{X_\tau', \tau \in (\alpha_n, \beta_n)\} = \sup \{X'_t, t \in (\alpha_n, \beta_n)\}$$

for every n, since by the definition of $l.$-th. sup,

$$l.\text{-th. sup } \{X'_\tau, \tau \in (a, b)\} \geq l.\text{-th. sup } \{X'_\tau, \tau \in (\alpha_n, \beta_n)\}$$

and yet the countable supremum,

$$\sup_n l.\text{-th. sup } \{X'_\tau, \tau \in (\alpha_n, \beta_n)\}$$

does satisfy (a) of Theorem 1 for $\Lambda = (a, b)$. For fixed n, every $t \in (\alpha_n, \beta_n)$ we have by (1) that $X'_t(\omega) \leq U_m(\omega)$ for all ω where U_m corresponds to (α_n, β_n) in the definition of the U_n above. For all $\omega \in \Omega$, $\sup \{X'_t(\omega), t \in (\alpha_n, \beta_n)\} \leq U_m(\omega)$. By means of (b) of Theorem 1 and the fact that $P[X_t = X'_t] = 1$ for every T,

we obtain

$$l.\text{-th. sup } \{X'_t, t \in (\alpha_n, \beta_n)\} = U_m \text{ a.s.},$$

and hence

$$\sup \{X'_t(\omega), t \in (\alpha_n, \beta_n)\} \leq l.\text{-th. sup } \{X'_t, t \in (\alpha_n, \beta_n)\} \text{ a.s.}$$

By conclusion (b) of Theorem 1 the reverse inequality is easily established. Q.E.D.

Definition: Let $\{X_t, t \in T\}$ be a separable stochastic process. For every open interval $(a, b) \subset T$ with rational endpoints there exists by conclusion (b) in Theorem 1 a sequence $\{t_n\} \subset (a, b)$ such that $\sup_n X_{t_n} = \sup \{X_t, t \in (a, b)\}$ a.s. and $\inf_n X_{t_n} = \inf \{X_t, t \in (a, b)\}$ a.s. Select one such sequence for every $(a, b) \subset T$ with rational endpoints; the union S of all such sequences for all such intervals (a, b) is obviously countable and is called a separating set for the separable stochastic process $\{X_t, t \in T\}$.

The very easy properties of separating sets for separable stochastic processes are left as exercises.

EXERCISES

1. If $\{X_\lambda, \lambda \in \Lambda\}$ is a set of random variables, and if Λ is countable, prove that

$$\sup \{X_\lambda, \lambda \in \Lambda\} = l.\text{-th. sup } \{X_\lambda, \lambda \in \Lambda\} \text{ a.s.}$$

2. Prove that if $\{X_t, t \in T\}$ is a separable stochastic process, if S is a separating set, and if S' is a countable set satisfying $S \subset S' \subset T$, then S' is a separating set.

3. Prove: A stochastic process $\{X_t, t \in T\}$ is separable if and only if there is a countable subset S of T such that for every open interval $(a, b) \subset T$,

$$\sup \{X_\tau \mid \tau \in S \cap (a, b)\} = \sup \{X_\tau \mid \tau \in (a, b)\} \text{ a.s.}$$

and

$$\inf \{X_\tau \mid \tau \in S \cap (a, b)\} = \inf \{X_\tau \mid \tau \in (a, b)\} \text{ a.s.}$$

4. Verify: Two stochastic processes $\{X_t, t \in T\}$ and $\{X'_t, t \in T\}$ that are defined over the same probability space and have the same index set T are equivalent if $P[X_t = X'_t] = 1$ for every $t \in T$.

5. Prove: If $\{X_t, t \in T\}$ is a separable stochastic process, and if S is a subset of T having properties of Exercise 3, then S is a separating set.

6. Prove: If S is a separating set for a separable stochastic process $\{X_t, t \in T\}$, then S is dense in T.

7. Prove: If $\{X_t, t \in [0, T]\}$ is a separable stochastic process which is continuous in probability, that is, $P[\,|\,X_\tau - X_t\,| \geq \epsilon\,] \to 0$ as $\tau \to t$ for all $\epsilon > 0$ and $t \in [0, T]$, then any countable dense subset of $[0, T]$ is a separating set.

8.3. Continuity and Nonrectifiability of Almost All Sample Functions of Separable Brownian Motion

One of the most important stochastic processes is the Brownian motion stochastic process that was introduced in Section 7.4. In this section we prove that a Brownian motion stochastic process is continuous with probability one, that the set of sample functions differentiable at a point t_0 is an event of probability zero, and that the set of all rectifiable sample functions over any bounded interval $I \subset [0, \infty)$ is an event of probability zero.

Lemma 1. If $\{X(t), t \in T\}$ is a separable stochastic process, and if I is a subinterval of T, then

$$\sup \{\,|\, X(t') - X(t'')\,|, t', t'' \in I\} \quad \text{and} \quad \sup \{\,|\, X(\tau) - X(t)\,|, \tau \in I\}$$

are both $\mathfrak{a}$-measurable.

Proof: We need only prove the lemma in the case that I is open. Separability implies that

$$\sup \{X(t') \mid t' \in I\} = l.\text{-th. sup } \{X(t'), t' \in I\} \text{ a.s.}$$

and

$$\inf \{X(t'') \mid t'' \in I\} = l.\text{-th. inf } \{X(t''), t'' \in I\} \text{ a.s.}$$

are both $\mathcal{Q}$-measurable, and hence

$$\sup \{X(t') \mid t' \in I\} - \inf \{X(t'') \mid t'' \in I\}$$

$$= \sup \{X(t') - X(t'') \mid t', t'' \in I\}$$

is $\mathcal{Q}$-measurable. Similarly $\inf \{X(t') - X(t'') \mid t', t'' \in I\}$ is $\mathcal{Q}$-measurable. Thus the conclusion that $\sup \{ \mid X(t') - X(t'') \mid, t', t'' \in I\}$ is $\mathcal{Q}$-measurable follows from the fact that if S is any set of real numbers, then

$$\sup \{ \mid s \mid, s \in S\} = \max \{\sup S, -\inf S\}.$$

A similar argument proves that $\sup \{ \mid X(\tau) - X(t) \mid, \tau \in I\}$ is $\mathcal{Q}$-measurable. Q.E.D.

Lemma 2. If $X(t)$ is a separable Brownian motion stochastic process, then for $T > 0$,

$$P[\sup \{X(t) - X(0), 0 \le t \le T\} \ge \lambda] \le \frac{\sigma}{\lambda} \sqrt{\frac{2T}{\pi}} \exp \left\{ \frac{-\lambda^2}{2\sigma^2 T} \right\}.$$

Proof: Because of separability we know that $S = \sup \{X(t) - X(0), 0 \le t \le T\}$ is an $\mathcal{Q}$-measurable function. Because of separability we know there exists a dense sequence $\{t_n\} \subset [0, T]$ such that

$$S = \sup_n \{X(t_n) - X(0)\} \text{ a.s.}$$

Let us denote $S_n = \max \{X(t_k) - X(0), 1 \le k \le n\}$. Then $S_n \to S$ a.s. Let us relabel $t_1, t_2, \cdots, t_n$ as $\tau_{n,1}, \tau_{n,2}, \cdots, \tau_{n,n}$ such that $\tau_{n,0} = 0 \le \tau_{n,1} < \tau_{n,2} < \cdots < \tau_{n,n} \le T$. Let us denote

$$X_{n,j} = X(\tau_{n,j}) - X(\tau_{n,j-1}), \qquad 1 \le j \le n,$$

and

$$S_{n,j} = X_{n,1} + \cdots + X_{n,j}.$$

Then $S_{n,k} = X(\tau_{n,k}) - X(0)$, and since

$$\mathcal{L}(S_{n,k} - S_{n,n}) = \mathcal{L}(X(\tau_{n,k}) - X(\tau_{n,n})) = \mathfrak{N}(0, (\tau_{n,n} - \tau_{n,k})\sigma^2),$$

then med $(S_{n,k} - S_{n,n}) = 0$. By Lemma 2 of Section 5.6

$$P[S_n \ge \epsilon] \le 2P[S_{n,n} \ge \epsilon] = 2P[X(\tau_{n,n}) \ge \epsilon]$$

$$= \sqrt{2/\pi\sigma^2 \tau_{n,n}} \int_\epsilon^\infty \exp \left(-\frac{x^2}{2\sigma^2 \tau_{n,n}} \right) dx.$$

But

$$\int_\epsilon^\infty \exp\left(-\frac{x^2}{2\sigma^2\tau_{n,n}}\right) dx \leq \frac{1}{\epsilon} \int_\epsilon^\infty x \exp\left(-\frac{x^2}{2\sigma^2\tau_{n,n}}\right) dx$$

$$= (\sigma^2\tau_{n,n}/\epsilon) \int_{\epsilon/\sqrt{\sigma^2\tau_{n,n}}}^\infty u \exp\left(-u^2/2\right) du$$

$$= (\sigma^2\tau_{n,n}/\epsilon) \exp\left(-\epsilon^2/2\sigma^2\tau_{n,n}\right),$$

or

$$P[S_n \geq \epsilon] \leq \sqrt{2\sigma^2\tau_{n,n}/\pi} \, (1/\epsilon) \exp\left(-\epsilon^2/2\sigma^2\tau_{n,n}\right).$$

Not only does $S_n \to S$ a.s., but $[S_n \geq \epsilon] \to [S \geq \epsilon]$. This implies that $P[S_n \geq \epsilon] \to P[S \geq \epsilon]$. Also $\tau_{n,n} \to T$. Thus

$$P[S \geq \epsilon] \leq \sqrt{2T/\pi} \, (\sigma/\epsilon) \exp\left(-\epsilon^2/2\sigma^2 T\right),$$

which proves the lemma.

Lemma 3. If $\{X(t), t \geq 0\}$ is a separable Brownian motion, then for every $\tau \geq \epsilon > 0$ and every $\lambda > 0$,

$$P[\sup \{ \, | \, X(t) - X(\tau) \, |, \, \tau - \epsilon \leq t \leq \tau + \epsilon\} \leq \lambda]$$

$$= (P[\sup \{ \, | \, X(t) - X(0) \, |, \, 0 \leq t \leq \epsilon\} \leq \lambda])^2.$$

Proof: The probabilities in the lemma are well-defined because of Lemma 1. Clearly

$$P[\sup \{ \, | \, X(t) - X(\tau) \, |, \, \tau - \epsilon \leq t \leq \tau + \epsilon\} \leq \lambda]$$

$$= P[\max \{U, V\} \leq \lambda],$$

where

$$V = \sup \{ \, | \, X(t) - X(\tau) \, |, \, \tau - \epsilon \leq t \leq \tau\}$$

and

$$U = \sup \{ \, | \, X(t) - X(\tau) \, |, \, \tau \leq t \leq \tau + \epsilon\}.$$

Let us denote $Y(t) = X(t + \tau) - X(\tau)$ and $Z(t) = X(\tau - t) - X(\tau)$. Then by routine verification, one sees that $\{Y(t), 0 \leq t \leq \epsilon\}$ and $\{Z(t),$

$0 \leq t \leq \epsilon\}$ are separable Brownian motion stochastic processes with Var $Y(t) =$ Var $Z(t) = \sigma^2 t$ for all $t \in [0, \epsilon]$. This implies that U and V are identically distributed. Now let $\{t_n\}$ be a countable and dense subset of $[0, \epsilon]$ so that $U = \sup_n | Y(t_n) |$ a.s. and $V = \sup_n | Z(t_n) |$ a.s. (Separability and Theorem 1 in Section 8.2 allow this.) For every positive integer n, let $\tau_{n,0} = 0$, and let $0 \leq \tau_{n,1} < \cdots < \tau_{n,n} \leq \epsilon$ be an ordering of $t_1, t_2, \cdots, t_n$. Denote

$$Y_j = Y(\tau_{n,j}) - Y(\tau_{n,j-1}) \qquad \text{and} \qquad Z_j = Z(\tau_{n,j}) - Z(\tau_{n,j-1})$$

for $1 \leq j \leq n$. Since $\{X(t), t \geq 0\}$ has independent increments, it follows that the random variables $Y_1, \cdots, Y_n, Z_1, \cdots, Z_n$ are independent. Let $y_1, \cdots, y_n, z_1, \cdots, z_n$ be any $2n$ real numbers. Because of independence

$$E \exp \{i \sum_{j=1}^{n} y_j Y_j + i \sum_{j=1}^{n} z_j Z_j\} = E \exp \{i \sum_{j=1}^{n} y_j Y_j\} E \exp \{i \sum_{j=1}^{n} z_j Z_j\}.$$

Let us make the following substitution:

$$y_1 = u_1 + u_2 + \cdots + u_n, \qquad y_2 = u_2 + \cdots + u_n, \cdots, y_n = u_n,$$

$$z_1 = v_1 + v_2 + \cdots + v_n, \qquad z_2 = v_2 + \cdots + v_n, \cdots, z_n = v_n.$$

We then obtain from the above identity that

$$E \exp \{i \sum_{j=1}^{n} u_j Y(\tau_{n,j}) + i \sum_{j=1}^{n} v_j Z(\tau_{n,j})\}$$

$$= E \exp \{i \sum_{j=1}^{n} u_j Y(\tau_{nj})\} E \exp \{i \sum_{j=1}^{n} v_j Z(\tau_{n,j})\}.$$

Thus by this factorization of a joint characteristic function we obtain (by Exercise 9 in Section 3.2) that the two vectors

$$(Y(\tau_{n,1}), \cdots, Y(\tau_{n,n})) \qquad \text{and} \qquad (Z(\tau_{n,1}), \cdots, Z(\tau_{n,n}))$$

are independent. If we define

$$U_n = \max \{ | Y(\tau_{n,j}) |, 1 \leq j \leq n\}$$

and

$$V_n = \max \{ | Z(\tau_{n,j}) |, 1 \leq j \leq n\},$$

then U_n and V_n are independent for every n. (This follows from Exercise

1 in Section 3.2.) Further $[U_n \leq \lambda] \to [U \leq \lambda]$ and $[V_n \leq \lambda] \to [V \leq \lambda]$. Hence

$$P[\max\{U, V\} \leq \lambda] = P([U \leq \lambda][V \leq \lambda])$$

$$= \lim_{n \to \infty} P([U_n \leq \lambda][V_n \leq \lambda])$$

$$= \lim_{n \to \infty} P[U_n \leq \lambda]P[V_n \leq \lambda]$$

$$= P[U \leq \lambda]P[V \leq \lambda] = (P[U \leq \lambda])^2. \quad \text{Q.E.D.}$$

Theorem 1. Let $\{X(t), 0 \leq t < \infty\}$ be a separable Brownian motion stochastic process with $\operatorname{Var} X(t) = \sigma^2 t$. Then the set of all continuous sample functions is an event of probability one. (The definition of sample function was given at the end of Section 8.1.)

Proof: For $N = 1, 2, \cdots$, denote

$$C_N = [\sup \{ \, | X(t) - X(j/N) \, | \, | \, | t - j/N \, | \leq 1/N\}$$
$$\leq 1/N^{1/4}, 1 \leq j \leq N^2].$$

Let $C^c = [C_N{}^c \text{ infinitely often}]$. We shall prove that $P(C^c) = 0$. Let us denote

$$A_{N,k} = [\sup \{ \, | X(t) - X(k/N) \, | \, | \, | t - k/N \, | \leq 1/N\} > N^{-1/4}]$$

and

$$B_N = [\sup \{ \, | X(t) - X(0) \, |, 0 \leq t \leq 1/N\} < N^{-1/4}].$$

By Boole's inequality and Lemmas 2 and 3 we have

$$P(C_N{}^c) = P(\bigcup_{k=1}^{N^2} A_{N,k}) \leq \sum_{k=1}^{N^2} P(A_{N,k})$$

$$= N^2 P(A_{N,1}) = N^2(1 - P(A_{N,1}^c))$$

$$\leq N^2(1 - P^2(B_N))$$

$$= N^2(1 - (1 - P(B_N^c))^2)$$

$$= N^2(2P(B_N^c) - P^2(B_N^c))$$

$$\leq 2N^2 P(B_N^c)$$

$$\leq \left(\frac{4\sqrt{2}\sigma}{\sqrt{\pi}}\right) N^{7/4} \exp\left\{\frac{-\sqrt{N}}{2\sigma^2}\right\}.$$

But $\sum P(C_N{}^c) < \infty$, and thus by the Borel-Cantelli lemma, $P(C^c) = 0$, or $P(C) = 1$. But C is the event that all but a finite number of C_N occur, that is, C_N occurs for all sufficiently large N with probability 1. Hence if $\omega \in C$, then for every t and every $\epsilon > 0$ there is an N_0 sufficiently large such that $N_0 > t + 1$, $2/N_0{}^{1/4} < \epsilon$ and $\omega \in C_{N_0}$. Then taking $\delta = 1/2N_0$ we have that if $| t - \tau | < \delta$, then, letting j_0 be such that $| j_0/N_0 - t | = \min \{ | j/N_0 - t | : j = 0, 1, 2, \cdots \}$, we have $| j_0/N_0 - \tau | < 1/N_0$, and so

$$| X(t, \omega) - X(\tau, \omega) |$$

$$\leq \left| X(t, \omega) - X\left(\frac{j_0}{N_0}, \omega\right) \right| + \left| X\left(\frac{j_0}{N_0}, \omega\right) - X(\tau, \omega) \right|$$

$$\leq \frac{1}{N_0{}^{1/4}} + \frac{1}{N_0{}^{1/4}} = \frac{2}{N_0{}^{1/4}} < \epsilon,$$

that is, $X(t, \omega)$ is continuous in t if $\omega \in C$. Q.E.D.

Note that Theorem 1 solves the second problem stated in Section 8.1 in the case when one wishes to fit the measure for Brownian motion over the space of all continuous functions over $[0, \infty)$ which vanish at 0.

Lemma 4. If $X_1, X_2, \cdots, X_n$ are independent random variables whose distribution functions are symmetric, if $S_j = X_1 + \cdots + X_j$, and if $\lambda > 0$ and $\epsilon > 0$, then

$$P[\max \{S_j \,|\, 1 \leq j \leq n\} \geq \lambda]$$

$$\geq 2P[S_n \geq \lambda + 2\epsilon] - 2 \sum_{j=1}^{n} P[X_j \geq \epsilon].$$

Proof: Let us define $N = \min \{ j \,|\, S_j \geq \lambda \}$. Then N is a random variable. One easily verifies that

$$[N = k][S_n - S_k < -\epsilon][X_k < \epsilon]$$

$$\subset [N = k][S_n < S_k - \epsilon][X_k < \epsilon][S_{k-1} < \lambda]$$

$$\subset [N = k][S_n < S_{k-1} + X_k - \epsilon][X_k - \epsilon < 0][S_{k-1} < \lambda]$$

$$\subset [N = k][S_n < S_{k-1}][S_{k-1} < \lambda]$$

$$\subset [N = k][S_n < \lambda],$$

or

$$[N = k][S_n < \lambda] \supset [N = k][S_n - S_k < -\epsilon][X_k \geq \epsilon]^c.$$

Taking probabilities we obtain

$$P([N = k][S_n < \lambda]) \geq P([N = k][S_n - S_k < -\epsilon]) - P[X_k \geq \epsilon].$$

Hence

$$P[\max \{S_j \mid 1 \leq j \leq n\} \geq \lambda, \, S_n < \lambda]$$

$$= \sum_{k=1}^{n-1} P([N = k][S_n < \lambda])$$

$$\geq \sum_{k=1}^{n-1} P([N = k][S_n - S_k < -\epsilon]) - \sum_{k=1}^{n-1} P[X_k \geq \epsilon].$$

Now $[N = k] = [S_1 < \lambda] \cdots [S_{k-1} < \lambda][S_k \geq \lambda]$ depends only on $X_1, \cdots, X_k$, and $[S_n - S_k < -\epsilon]$ depends on $X_{k+1}, \cdots, X_n$. Hence the two events are independent. Further, since $X_1, \cdots, X_n$ are independent and have symmetric distributions, it follows (by a characteristic function argument) that the distribution of $S_n - S_k = X_{k+1} + \cdots + X_n$ is symmetric, and thus

$$P([N = k][S_n - S_k < -\epsilon])$$

$$= P[N = k]P[S_n - S_k < -\epsilon]$$

$$= P[N = k]P[S_n - S_k > \epsilon]$$

$$= P([N = k][S_n - S_k > \epsilon]).$$

Hence we obtain

(1) $P[\max \{S_j \mid 1 \leq j \leq n\} \geq \lambda, \, S_n < \lambda]$

$$\geq \sum_{k=1}^{n-1} P([N = k][S_n - S_k > \epsilon]) - \sum_{k=1}^{n-1} P[X_k \geq \epsilon].$$

One easily verifies that

$$[N = k][S_n \geq \lambda + 2\epsilon][X_k < \epsilon]$$

$$\subset [N = k][S_{k-1} < \lambda][S_k \geq \lambda][S_n \geq \lambda + 2\epsilon][X_k < \epsilon],$$

and since

$$[S_{k-1} < \lambda][S_k \geq \lambda][X_k < \epsilon] \subset [S_k < \lambda + \epsilon],$$

we have

$$[N = k][S_n \geq \lambda + 2\epsilon][X_k < \epsilon]$$
$$\subset [N = k][S_n \geq \lambda + 2\epsilon][S_k < \lambda + \epsilon]$$
$$\subset [N = k][S_n - S_k > \epsilon].$$

Taking probabilities we obtain

$$P([N = k][S_n - S_k > \epsilon])$$
$$\geq P([N = k][S_n \geq \lambda + 2\epsilon]) - P[X_k \geq \epsilon].$$

Substituting this inequality into (1) we obtain

$$(2) \quad P([\max \{S_j \mid 1 \leq j \leq n\} \geq \lambda][S_n < \lambda])$$
$$\geq \sum_{k=1}^{n-1} P([N = k][S_n \geq \lambda + 2\epsilon]) - 2 \sum_{k=1}^{n-1} P[X_k \geq \epsilon].$$

Since

$$[S_n \geq \lambda + 2\epsilon] \subset [\max \{S_j \mid 1 \leq j \leq n\} \geq \lambda][S_n \geq \lambda],$$

we have

$$(3) \quad P([\max \{S_j \mid 1 \leq j \leq n\} \geq \lambda][S_n \geq \lambda]) \geq P[S_n \geq \lambda + 2\epsilon].$$

Also, one can verify that

$$\bigcup_{k=1}^{n-1} [N = k][S_n \geq \lambda + 2\epsilon] \supset [S_n \geq \lambda + 2\epsilon][X_n < \epsilon],$$

and taking probabilities we find that

$$(4) \quad \sum_{k=1}^{n-1} P([N = k][S_n \geq \lambda + 2\epsilon]) \geq P[S_n \geq \lambda + 2\epsilon] - P[X_n \geq \epsilon].$$

Adding both sides of (2) and (3) and substituting (4) into (2), we obtain the lemma. Q.E.D.

Theorem 2. If $X(t)$, $0 \leq t < \infty$, is a separable Brownian motion stochastic process, then, for $\lambda \geq 0$,

$$P[\sup \{X(t) - X(0) \mid t \in [0, T]\} \geq \lambda]$$
$$= 2P[X(T) - X(0) \geq \lambda] = \sqrt{2/\pi\sigma^2 T} \int_{\lambda}^{\infty} \exp\left(-\frac{t^2}{2\sigma^2 T}\right) dt.$$

Proof: Separability and Inequality (1) of Lemma 2 in Section 5.6 easily imply

$$P[\sup \{X(t) \mid 0 \le t \le T\} \ge \lambda] \le 2P[X(T) \ge \lambda].$$

In order to prove the reverse inequality, let $\{t_n\}$ be a sequence in $[0, T]$ such that $t_1 = T$, such that

$$\sup \{X(t) \mid 0 \le t \le T\} = \sup \{X(t_n) \mid n = 1, 2, \cdots\} \text{ a.s.},$$

and such that $\{t_n\}$ is dense in $[0, T]$; this is possible because of separability. Let $0 \le t_{n,1} < t_{n,2} < \cdots < t_{n,n} \le T$ be $t_1, \cdots, t_n$ reordered, and denote $t_{n,0} = 0$. Let us denote $X_{n,j} = X(t_{n,j}) - X(t_{n,j-1})$ and $S_{n.j} = X(t_{n,j}) = X_{n,1} + \cdots + X_{n,j}$. By Lemma 4,

$$P[\max \{S_{n,j} \mid 1 \le j \le n\} \ge \lambda]$$

$$\ge 2P[S_{n,n} \ge \lambda + 2\epsilon] - 2 \sum_{j=1}^{n} P[X_{n,j} \ge \epsilon],$$

where $\epsilon > 0$ is arbitrary. But

$$P[\max \{S_{n,j} \mid 1 \le j \le n\} \ge \lambda] \to P[\sup \{X(t) \mid 0 \le t \le T\} \ge \lambda]$$

as $n \to \infty$ at all $\lambda > 0$ at which the limiting distribution is continuous. Also, for all $\lambda > 0$, $\epsilon > 0$,

$$P[S_{n,n} \ge \lambda + 2\epsilon] = P[X(T) \ge \lambda + 2\epsilon]$$

for all n. We next observe that

$$\int_{\lambda}^{\infty} \exp\left(-\frac{x^2}{2}\right) dx \le \frac{1}{\lambda} \int_{\lambda}^{\infty} x \exp\left(-\frac{x^2}{2}\right) dx = \frac{1}{\lambda} \exp\left(-\frac{\lambda^2}{2}\right).$$

Let us denote $\sigma_j^2 = (t_{n,j} - t_{n,j-1})\sigma^2$. Then by this last inequality we obtain

$$P[X_{n,j} \ge \epsilon] = (2\pi\sigma_j^2)^{-1/2} \int_{\epsilon}^{\infty} \exp\left(-\frac{x^2}{2\sigma_j^2}\right) dx$$

$$= (2\pi)^{-1/2} \int_{\epsilon/\sigma_j}^{\infty} \exp\left(-\frac{t^2}{2}\right) dt$$

$$\le (2\pi)^{-1/2} \left(\frac{\sigma_j}{\epsilon}\right) \exp\left(-\frac{\epsilon^2}{2\sigma_j^2}\right).$$

By Cauchy's inequality,

$$\left(\sum_{j=1}^{n} P[X_{n,j} \geq \epsilon]\right)^2 \leq n \sum_{j=1}^{n} (P[X_{n,j} \geq \epsilon])^2$$

$$\leq \frac{n\sigma^2 T}{2\pi\epsilon^2} \max\left\{\exp\left(-\frac{\epsilon^2}{\sigma_j^2}\right) \middle| 1 \leq j \leq n\right\}.$$

Since $\{t_n\}$ are dense in $[0, T]$ they may be ordered so that

$$t_{n,j} - t_{n,j-1} < 2T/n \quad \text{for} \quad 1 \leq j \leq n, n = 1, 2, \cdots.$$

If we denote $K = \sigma^2 T/2\pi\epsilon^2$, then

$$\left(\sum_{j=1}^{n} P[X_{n,j} \geq \epsilon]\right)^2 \leq Kn \exp\left(-\frac{n\epsilon^2}{2T\sigma^2}\right) \to 0 \qquad \text{as} \quad n \to \infty.$$

If we let $n \to \infty$ and then $\epsilon \to 0$ we get the reverse inequality at all but a countable set of discontinuity values of λ. However, if two distribution functions are equal except at a countable set of points, then they are identically equal. Q.E.D.

Theorem 3. Let $\{X(t), 0 \leq t < \infty\}$ be a separable Brownian motion stochastic process. Then at each fixed t_0 the probability that $X(t)$ is differentiable is zero, and in particular

$$P\left[\limsup_{t \to t_0} \frac{X(t) - X(t_0)}{t - t_0} = \infty\right] = 1.$$

Proof: Let $\lambda > 0$ be an arbitrary (large) number. By Theorem 2,

$$P\left[\sup_{t_0 < t < t_0 + \delta} \frac{X(t) - X(t_0)}{t - t_0} \geq \lambda\right]$$

$$\geq P[\sup_{t_0 < t < t_0 + \delta} (X(t) - X(t_0)) \geq \lambda\delta]$$

$$= 2P[X(t_0 + \delta) - X(t_0) \geq \lambda\delta]$$

$$= \sqrt{2/\pi\sigma^2\delta} \int_{\lambda\delta}^{\infty} \exp\left(-\frac{t^2}{2\sigma^2\delta}\right) dt$$

$$= \sqrt{2/\pi} \int_{\lambda\sqrt{\delta/\sigma^2}}^{\infty} \exp\left(-\frac{t^2}{2}\right) dt \to 1 \qquad \text{as} \quad \delta \to 0.$$

Because of Lemma 3, we obtain from the above inequality and limit that

$$P\left[\sup\left\{\frac{X(t)-X(t_0)}{t-t_0}\middle| t_0-\delta<t<t_0+\delta, t\neq t_0\right\}\geq\lambda\right]\to 1$$

as $\delta\to 0$. From this we easily obtain the conclusion of the theorem (see Exercise 5).

Theorem 4. If $X(t)$, $0\leq t<\infty$, is a separable Brownian motion stochastic process, then the set of all nonrectifiable sample functions over the bounded interval $[0, T]$ is an event of probability one.

Proof: Let $0=t_{n,0}<t_{n,1}<\cdots<t_{n,n}=T$ be such that if $\mathcal{P}_n=[t_{n,0},\cdots,t_{n,n}]$, then $\mathcal{P}_n\subset\mathcal{P}_{n+1}$ for all n, and $\bigcup_{n=1}^{\infty}\mathcal{P}_n$ is dense in $[0, T]$. Then if we denote

$$S_n{}^2=\sum_{j=1}^{n}(X(t_{n,j})-X(t_{n,j-1}))^2,$$

we have

$$S_n{}^2\leq\max\{|X(t_{n,j})-X(t_{n,j-1})|, 1\leq j\leq n\}$$

$$\cdot\sum_{k=1}^{n}|X(t_{n,k})-X(t_{n,k-1})|.$$

By Theorem 2 in Section 7.4, $S_n{}^2\to\sigma^2 T$ a.s. But by Theorem 1,

$$\max\{|X(t_{n,j})-X(t_{n,j-1})|, 1\leq j\leq n\}\to 0\text{ a.s.}$$

as $n\to\infty$. Hence, since $0<\sigma^2 T<\infty$,

$$\sum_{k=1}^{n}|X(t_{n,k})-X(t_{n,k-1})|\to\infty\text{ a.s.}$$

as $n\to\infty$, which is the conclusion of the theorem.

EXERCISES

1. Prove: if $\{X(t), 0\leq t<\infty\}$ is a separable Brownian motion stochastic process, then for $\tau>\epsilon>0$ and $s\geq 0$ fixed,

$$\mathfrak{L}(\sup\{X(t)-X(s)\mid s\leq t\leq s+\epsilon\})$$

$$=\mathfrak{L}(-\inf\{X(t)-X(s)\mid s\leq t\leq s+\epsilon\})$$

$$=\mathfrak{L}(\sup\{X(u)-X(\tau)\mid\tau-\epsilon\leq u\leq\tau\}).$$

2. Prove that if $X_1,\cdots,X_n$ are independent random variables, each

with a symmetric distribution function, then the distribution function of $X_1 + \cdots + X_n$ is symmetric.

3. Prove: if $X(t), 0 \le t < \infty$, is a separable Brownian motion stochastic process, then

$$P\left[\liminf_{t \to t_0} \frac{X(t) - X(t_0)}{t - t_0} = -\infty\right] = 1.$$

4. Let $\{X(t),\ 0 \le t < \infty\}$ be a stochastic process with independent increments (that is, if $0 = t_{n,0} < \cdots < t_{n,n}$, then

$$X(t_{n,1}) - X(t_{n,0}), \cdots, X(t_{n,n}) - X(t_{n,n-1})$$

are independent random variables) and such that $\mathcal{L}(X(t + h) - X(t))$ does not depend on t. Prove that if the set of all continuous sample functions of $X(t), 0 \le t < \infty$, is an event of probability one, then the stochastic process $X(t) - X(0), 0 \le t < \infty$, is a separable Brownian motion. (Hint: see Theorem 2 in Section 6.6.)

5. Let $Z(t), t \ge 0$, be a separable stochastic process. Prove that if

$$P[\sup_{|t - t_0| < \delta} Z_t \ge \lambda] \to 1 \qquad \text{as} \quad \delta \to 0,$$

then

$$P[\limsup_{t \to t_0} Z_t \ge \lambda] = 1.$$

6. Prove that $\sum_{N=1}^{\infty} N^{7/4} \exp\{-N^{1/2}/2\sigma^2\} < \infty$.

8.4. The Law of the Iterated Logarithm for Separable Brownian Motion

In Section 5.6 the law of the iterated logarithm was proved for sums of independent random variables. This section is devoted to a proof of the following theorem, which is the law of the iterated logarithm for a separable Brownian motion.

Theorem 1. If $\{X(t), 0 \le t < \infty\}$ is a separable Brownian motion with Var $X(t) = t$, then

(1)
$$P\left[\limsup_{t \to \infty} \frac{X(t)}{\sqrt{2t \log \log t}} = 1\right] = 1,$$

and

(2)
$$P\left[\limsup_{t \to 0} \frac{X(t)}{\sqrt{2t \log \log (1/t)}} = 1\right] = 1.$$

Proof: We first prove that the probabilities given in the statement of the theorem are well-defined. Indeed, because of separability, then by Theorem 1 in Section 8.3 almost all sample functions of $\{X(t), t \geq 0\}$ are continuous. Since $\sqrt{2t \log \log t}$ is continuous and bounded away from zero for $t \geq 3$, then for any interval $(a, b) \subset (3, \infty)$,

$$\sup_{t \in (a,b)} \frac{X(t)}{\sqrt{2t \log \log t}} = \sup_n \frac{X(t_n)}{\sqrt{2t_n \log \log t_n}} \quad \text{a.s.}$$

for any sequence $\{t_n\}$ which is dense in (a, b). Thus, such suprema are random variables. In order to prove (1) it is sufficient to prove that for every $\epsilon > 0$,

$$(3) \qquad P\left[\limsup_{t \to \infty} \frac{X(t)}{\sqrt{2t \log \log t}} > 1 + \epsilon\right] = 0,$$

and

$$(4) \qquad P\left[\limsup_{t \to \infty} \frac{X(t)}{\sqrt{2t \log \log t}} > 1 - \epsilon\right] = 1.$$

We first prove (3). It is sufficient to show that for $\epsilon > 0$ there exists a t_0 such that

$$P[X(t) \leq (1 + \epsilon)\sqrt{2t \log \log t} \quad \text{for all} \quad t \geq t_0] = 1.$$

Let $a > 1$ and let us denote

$$A_k = [\sup \{X(t) \mid 0 \leq t \leq a^k\} > (1 + \epsilon/2)\sqrt{2a^k \log \log a^k}].$$

Then by Theorem 2 in Section 8.4,

$$P(A_k) = \sqrt{2/\pi a^k} \int_{(1+\epsilon/2)\sqrt{2a^k \log \log a^k}}^{\infty} \exp\left(-\frac{y^2}{2a^k}\right) dy$$

$$\leq \sqrt{2/\pi a^k} \int_{(1+\epsilon/2)\sqrt{2a^k \log \log a^k}}^{\infty} \frac{y}{(1+\epsilon/2)\sqrt{2a^k \log \log a^k}}$$

$$\times \exp\left(-\frac{y^2}{2a^k}\right) dy$$

$$= \frac{1}{(1+\epsilon/2)\sqrt{\pi \log \log a^k}} \exp\{-(1+\epsilon/2)^2(\log \log a^k)\}$$

$$= K_k/(k \log a)^{(1+\epsilon/2)^2},$$

where

$$K_k = \frac{1}{(1 + \epsilon/2)\{\pi(\log k + \log \log a)\}^{1/2}}.$$

Hence $\sum_{k=1}^{\infty} P(A_k) < \infty$, so by the Borel-Cantelli lemma,

$$P[A_k \text{ infinitely often}] = 0.$$

Hence with probability one, for all sufficiently large k, $(k \geq k_0(\omega))$, and for all $t \in [a^{k-1}, a^k]$,

$$X(t) \leq (1 + \epsilon/2) \sqrt{2a^k \log \log a^k}$$
$$\leq (1 + \epsilon/2) \sqrt{2t \log \log t} \sqrt{(2a^k \log \log a^k)/(2a^{k-1} \log \log a^{k-1})}.$$

Preselect $a > 1$ so that

$$(1 + \epsilon/2) \sqrt{a} \left\{ \frac{\log k + \log \log a}{\log (k - 1) + \log \log a} \right\}^{1/2} < 1 + \frac{3\epsilon}{4}$$

for all $k \geq 2$. Then for t sufficiently large (depending on ω),

$$X(t) \leq (1 + \epsilon) \sqrt{2t \log \log t}$$

with probability one. If we had taken $a < 1$ but sufficiently close to 1 we could have proved in the same manner the first half of the local law, namely,

(3′) $$P\left[\limsup_{t \to 0} \frac{X(t)}{\sqrt{2t \log |\log t|}} > 1 + \epsilon \right] = 0$$

for every $\epsilon > 0$.

Now we wish to prove (4). Let us again take $a > 1$ (but this time a will be required to be large). Let us denote

$$B_k = [X(a^k) - X(a^{k-1}) \geq (1 - \epsilon/2) \sqrt{2a^k \log \log a^k}],$$

where $0 < \epsilon < 1$, but otherwise ϵ being arbitrary. Since $X(t)$ has independent increments (by the definition of Brownian motion), it follows that the events $\{B_k\}$ are independent. Since

$$\mathcal{L}(X(a^k) - X(a^{k-1})) = \mathfrak{N}(0, a^k - a^{k-1}),$$

then

$$P(B_k) = (2\pi(a^k - a^{k-1}))^{-1/2}$$

$$\times \int_{(1-\epsilon/2)\sqrt{2a^k \log \log a^k}}^{\infty} \exp\left[-y^2/2(a^k - a^{k-1})\right] dy.$$

Let us make the substitution $z = y/\sqrt{a^k - a^{k-1}} = y/\sqrt{a^k(1 - 1/a)}$. Hence if we denote

$$c_k = (1 - \epsilon/2)\sqrt{(2\log\log a^k)/(1 - 1/a)} \quad \text{and} \quad d_k = c_k + 1,$$

then

$$P(B_k) = (2\pi)^{-1/2}\int_{c_k}^{\infty} \exp(-z^2/2)\,dz$$

$$> (2\pi)^{-1/2}\int_{c_k}^{d_k} \exp(-z^2/2)\,dz$$

$$> (2\pi)^{-1/2}\exp\{-d_k^2/2\}.$$

Let us denote $\kappa = (1 - \epsilon/2)/\sqrt{1 - 1/a}$. If we take sufficiently large a, then $1 - 1/a$ is close to 1, and thus for large a, $\kappa < 1 - \epsilon/4$. Now

$$\tfrac{1}{2}\{\kappa\sqrt{2\log\log a^k} + 1\}^2 = \kappa^2(\log k + \log\log a)$$
$$+ \kappa\sqrt{2\log\log a^k} + \tfrac{1}{2}.$$

Since for all x sufficiently large, $K\sqrt{x} < \theta x$ for any $K > 0$, $\theta > 0$, we have then for all sufficiently large values of k that

$$\kappa\sqrt{2\log\log a^k} < \tfrac{1}{2}(1 - \kappa^2)\log\log a^k.$$

Hence for sufficiently large k,

$$\tfrac{1}{2}\{\kappa\sqrt{2\log\log a^k} + 1\}^2 < \tfrac{1}{2}(\kappa^2 + 1)(\log k + \log\log a).$$

From this we obtain

$$\exp\tfrac{1}{2}\{\kappa\sqrt{2\log\log a^k} + 1\}^2$$
$$< \exp\{((\kappa^2 + 1)/2)(\log k + \log\log a)\}$$
$$= (\log a)^{(\kappa^2+1)/2}k^{(\kappa^2+1)/2},$$

or

$$P(B_k) > \left(\frac{(\log a)^{\kappa^2+1}}{2\pi}\right)^{1/2}k^{-\beta},$$

where $0 < \beta = (\kappa^2 + 1)/2 < 1$. Thus $\sum P(B_n) = \infty$, and by the Borel lemmas (Theorem 2 of Section 3.3) we have $P[B_n \text{ i.o.}] = 1$. Since the distribution of stochastic process $\{X(t),\, t \geq 0\}$ is symmetric, we obtain from (3), which we have already proved, that for any $\delta > 0$,

$$P[-X(a^{k-1}) \leq (1 + \delta)\sqrt{2a^{k-1}\log\log a^{k-1}} \quad \text{for } \textit{all} \text{ large } k] = 1.$$

Hence

$$P[X(a^k) > (1 - \epsilon/2)\sqrt{2a^k \log\log a^k}$$
$$- (1 + \delta)\sqrt{2a^{k-1}\log\log a^{k-1}} \quad \text{for infinitely many } k] = 1.$$

But

$$(1 - \epsilon/2)\sqrt{2a^k \log\log a^k} - (1 + \delta)\sqrt{2a^{k-1}\log\log a^{k-1}}$$
$$= \sqrt{2a^k \log\log a^k}\{1 - \epsilon/2 - J\},$$

where

$$J = \frac{1+\delta}{\sqrt{a}}\sqrt{\frac{\log\log a^{k-1}}{\log\log a^k}}.$$

Again, preselect a so large so that $J < \epsilon/2$ for all $k \geq 2$. Then

$$P[X(a^k) > (1 - \epsilon)\sqrt{2a^k \log\log a^k} \text{ i.o.}] = 1,$$

or

$$P\left[\limsup_{k \to \infty} \frac{X(a^k)}{\sqrt{2a^k \log\log a^k}} > 1 - \epsilon\right] = 1.$$

Since

$$\left[\limsup_{k \to \infty} \frac{X(a^k)}{\sqrt{2a^k \log\log a^k}} > 1 - \epsilon\right]$$
$$\subset \left[\limsup_{t \to \infty} \frac{X(t)}{\sqrt{2t \log\log t}} > 1 - \epsilon\right],$$

we have established (4). If one selects $0 < a < 1$, with a close to zero, then in a manner similar to the proof of (4) one can establish

(4′) $$P\left[\limsup_{t \to \infty} \frac{X(t)}{\sqrt{2t \log |\log t|}} > 1 - \epsilon\right] = 1,$$

which together with (3′) establishes (2). Q.E.D.

EXERCISE

1. Carry through the details of the proof of (3′) and (4′).

Suggested Reading

Books in Probability Theory:

[1] J. L. Doob, "Stochastic Processes," Wiley, New York, 1953.
[2] W. Feller, "An Introduction to Probability Theory and Its Applications," Wiley, New York, Vol. I, 2nd ed., 1957; Vol. II, 1966.
[3] B. V. Gnedenko, "The Theory of Probability" (translated by B. D. Seckler), Chelsea, New York, 1962.
[4] B. V. Gnedenko and A. N. Kolmogorov, "Limit Distributions for Sums of Independent Random Variables" (translated by K. L. Chung), Addision-Wesley, Cambridge, Mass., 1954.
[5] M. Kac, "Statistical Independence in Probability, Analysis, and Number Theory," *Carus Mathematical Monographs* No. 12. Am. Math. Assoc., Providence, Rhode Island, 1959.
[6] A. N. Kolmogorov, "Foundations of Probability" (translated by Nathan Morrison), Chelsea, New York, 1950.
[7] P. Lévy, "Théorie de l'Addition des Variables Aléatoires," Gauthier-Villars, Paris, 1937.
[8] M. Loève, "Probability Theory," 3rd ed., Van Nostrand, Princeton, 1963.
[9] E. Lukacs, "Characteristic Functions," Hafner, New York, 1960.
[10] J. Neveu, "Mathematical Foundations of the Calculus of Probability," Holden-Day, San Francisco, 1965.
[11] M. Rosenblatt, "Random Processes," Oxford Univ. Press, New York, 1962.

References to Measure and Integration:

[1] P. R. Halmos, "Measure Theory," Van Nostrand, Princeton, 1950.
[2] E. Hewitt and K. Stromberg, "Real and Abstract Analysis," Springer-Verlag, New York, 1965.
[3] M. E. Munroe, "Introduction to Measure and Integration," Addison-Wesley, Cambridge, Mass., 1953.
[4] H. L. Royden, "Real Analysis," Macmillan, New York, 1963.
[5] S. Saks, "Theory of the Integral," Hafner, New York, 1936.

Journals in which many research papers in probability appear:

[1] The Annals of Mathematical Statistics.
[2] Illinois Journal of Mathematics.
[3] Theory of Probability and Its Applications. (This is a SIAM translation of a Russian journal.)
[4] Transactions of the American Mathematical Society.
[5] Zeitschrift der Wahrscheinlichkeitstheorie.

Index

Absolutely continuous 17, 27
Almost always 3
Almost sure (a.s.) convergence 99, of a series 107

Bayes' rule 8
Beta distribution 18
Beta function 18
Binomial distribution 19, 38, 45
Boole's inequality 6
Borel lemmas 72
Borel-Cantelli lemma 70
Brownian motion, definition of 239, sample functions properties of 254 ff

Cantor distribution 20, 23, 38, 115
Cantor ternary set 20
Cauchy convergent in probability 100
Cauchy distribution 18, 38, 47
Cell in $E^{(n)}$ 24, closed 77
Centering constants 178
"Central limit theorem" 205
Central moment 39
Characteristic function, definition of 42, joint 42, logarithm of a 92
Chebishev's inequality 40
Closed cell 77
Closing random variable 233
Closing random variable, nearest 233
Complex-valued random variable, definition of a 39, expectation of a 39
Concentration function 67, 88
Conditional expectation, definition of 210, most frequently used properties of 212, 213
Conditional probability 7
Cont F 77

Continuity theorem 88
Convergence, almost sure 99, complete 83, 85
Convergence in law 99
Convergence in probability 100
Convergence in rth mean 105
Convergence in the wide sense 78
Convergence to the Poisson distribution 205
Convex function 216
Convolution 21
Consistency condition 30
Continuous singular 20, 28
Correlation coefficient 41

Daniell-Kolmogorov theorem 30
Density 17
Discrete 28
Discrete distribution function 18
Distribution function, definition of 15, point of increase of a 22
Distribution of a set of infinitely many random variables 29
Double sequence of random variables which are row-wise independent 167

Egorov's theorem 101, 231, 232
Elementary event 1
Empirical distribution function 127
Equicontinuous 94
Event, definition of 2, elementary 1, impossible 2, indicator of an 11, sure 1
Expectation, definition of 34, key theorem for 35, computations of 37
Extension theorem for measures 238

Fatou's lemma, conditional form of 216

Gamma distribution 18
Gaussian distribution 17
Geometric distribution 20
Glivenko-Cantelli theorem 127

Helly compactness theorem 83
Helly-Bray theorem 84

Impossible event 2
Independence, best criterion for determining 63
Independence of classes of events 60
Independent events 58, class of 58
Independent random variables 61
Independent increments 239
Indicator of an event 11
Infinitely divisible characteristic function 148, canonical representation of a 153
Infinitely divisible distribution function 147
Infinitely divisible random variable 148
"Infinitely often" (i.o.) 3
Infinitesimal system 147, 167
Integrability, uniform 231

Jensen's inequality 216, conditional form of 216
Jump 18

Kolmogorov-Daniell theorem 30
Kolmogorov's inequalities 107
Kronecker's lemma 122

l.-th inf 250
l.-th sup 250
Lattice-theoretic infimum 250
Lattice-theoretic supremum 250
Law, definition of 99, limit 178
Law of the iterated logarithm 140
Law of the iterated logarithm for a separable Brownian motion 265
Lebesgue dominated convergence theorem, conditional form of 216
Lebesgue monotone convergence theorem, conditional form of 215
Lévy inequalities 137

Lévy representation 160
Lévy spectral function 184
lim 3
lim inf 3
lim sup 2
Limit law 178
Logarithm of a characteristic function 92

Marginal or marginal distribution 24
Martingale 220, reverse 221
"Martingale convergence theorem" 229
Median 115
Moments of a random variable 39
Multiplication rule 7
Multivariate distribution function 23

$\mathfrak{N}(\mu, \sigma^2)$ 17
Negative exponential distribution 18
Normal distribution 17, 37, 45, 148
Normal numbers 126

Orthogonality of two probability measures 244

$\mathcal{P}(\lambda)$ 20
Poisson component intensity function 163
Poisson distribution 19, 38, 45
Population 126
Positive definite 56
Power series with random coefficients 75
Probability 5
Probability space 7
Pseudometric space 9

Rademacher functions 68, 114
Radon-Nikodym theorem 209
Random variable, definition of a 9, discrete 12, n-dimensional 10, nth moment of a 39, vector 10
Random walk 146
Reverse martingale 221
Reverse submartingale 221

Sample 126
Sample function 248

Separable stochastic process 250
Separating set for a separable stochastic
   process 253
Sequence of centered sums of independ-
   ent random variables from an infini-
   tesimal system 178
Sigma field 2
Sigma field generated by a field 3
Sigma field induced or generated or de-
   termined by a set of random variables
   12
Stochastic process 245
Strong law of large numbers 124
Submartingale 221
Submartingale, reverse 221
Sure event 1
Symmetric difference 73
Symmetric distribution function 53

Tail sigma field 73
Three series theorem 113
Total probabilities, theorem of 8
Truncation 71

Uniform distribution 17, 20, 48
Uniform equicontinuity 94
Uniform integrability 231
Uniqueness of the extension theorem 242
Uniqueness theorem 51
Unit interval probability space 8, 60,
   69, 220, 228, 246, 248
Up-crossings, number of 225

Variance 39

Zero-or-one law 74